Late 17th- to 19th-century burial and earlier occupation at All Saints, Chelsea Old Church, Royal Borough of Kensington and Chelsea

GW01186382

MoLAS Archaeology Studies Series

Late 17th- to 19th-century burial and earlier occupation at All Saints, Chelsea Old Church, Royal Borough of Kensington and Chelsea

Robert Cowie, Jelena Bekvalac and

Tania Kausmally

MoLAS Archaeology Studies Series 18

Museum of London Archaeology Service

Published by the Museum of London Archaeology Service

Copyright © Museum of London 2008

A CIP catalogue record for this book is available from the British Library

Production and series design by Tracy Wellman
Typesetting and design by Sue Cawood
Reprographics by Andy Chopping
Copy editing by Wendy Sherlock
Series editing by Sue Hirst/Susan M Wright
Post-excavation and series management by Peter Rowsome

Printed by the Lavenham Press

Front cover: engraving showing Chelsea Old Church from the south-east
(c 1750) (Guildhall Library, City of London)

CONTRIBUTORS

Principal author (stratigraphy and historical background)	Robert Cowie
Principal authors (human bone)	Jelena Bekvalac, Tania Kausmally
Genealogical research	Robert Cowie, Richard Hewett, with Adrian Miles
Flint	Philippa Bradley
Building material	Susan Pringle
Clay tobacco pipes	Kieron Heard
Roman pottery	Rupert Featherby
Medieval and post-medieval pottery	Lyn Blackmore
Non-ceramic finds	Nicola Powell
Coffin furniture	Adrian Miles
Graphics	Sandra Rowntree (finds); Sophie Lamb, Peter Hart-Allison, Carlos Lemos (plans)
Photography	Andy Chopping, Maggie Cox
Project managers	Niall Roycroft, David Bowsher
Editor	David Bowsher

CONTENTS

FIGURES

TABLES

SUMMARY

Archaeological excavations undertaken in 2000 at 2–4 Old Church Street, Chelsea, revealed evidence from the prehistoric period to the 19th century. The site was located immediately north of All Saints, Chelsea Old Church, which was extensively rebuilt after it was bombed in 1941.

Prehistoric activity was indicated by at least nine struck flints, comprising waste flakes and a mesolithic burin. Several Roman features including pits, ditches and the possible remains of a rectangular timber building provided rare evidence for rural settlement in Londinium's immediate hinterland. The main phase of occupation appeared to be in the 3rd century AD.

In the medieval and Tudor periods, the site lay in the grounds of a manor house, which was probably situated to the north-east of the medieval church. The earliest medieval activity on the site is represented by a small quantity of residual pottery dated to c 1050–1150. A ditch and more than two dozen pits, mainly dated to the 13th and 14th centuries, were probably associated with seigneurial gardens, and included a row of pits that may have been used for planting. Building material from the site was fragmentary; some may have come from the medieval church.

From the late 17th century the northern half of the site was occupied by houses fronting on to Church Lane (now Old Church Street) and their back gardens and yards. The remains of two houses were found. Pits, wells and brick-lined cesspits, ranging in date from the 16th to 19th centuries, were scattered across the former back gardens. These mainly produced household and garden refuse, although the presence of crucibles in one pit suggests that there may have been small-scale fine metalworking in the area. The presence of red 'Tudor' brick, and such luxuries as polychrome tin-glazed floor tiles and a stove tile accord with documentary evidence for high-status buildings in the locality during the Tudor and Stuart periods.

In the late 17th century the south-east corner of the site was extensively quarried for sand. Towards the end of the 17th century a brick wall was built across the middle of the site to mark the northern boundary of the churchyard. Earth was then dumped in the churchyard on the south side of the wall (including the previously quarried area) to raise the ground level. The northern part of the churchyard began to be used as a burial ground in about 1700. It contained at least ten rows of graves dated to the 18th century and first half of the 19th century. Two rows lay within Petyt House (a school and vestry that was built in the north-west corner of the churchyard in 1707). Many of the graves contained several stacked burials. There were also two burial vaults and two brick-lined graves. Most coffins were wooden, although there were also nine lead-lined coffins. Nineteen burials have been positively identified from coffin plates, including two members of the Hand family who ran the famous Chelsea Bun House. A further two individuals have been provisionally identified.

The skeletons of 290 people, including two foetuses, were recovered, 198 of which were selected for detailed recording as part of the Wellcome Osteological Research Database. Most were chosen because they were relatively complete, but the recorded sample also included (irrespective of completeness) the remains of 25 individuals for which biographical information had been obtained from coffin plates and other sources.

Males and females were almost equally represented in the recorded sample; only 16.7% were subadults. Many of the adults survived into old age. Of the 22 people whose age was established from coffin plates 13 were aged 60 or over, and 4 were octogenarians. The study of this small subsample highlighted problems with current methods of establishing age-at-death from skeletal remains, which tend to underestimate the age of older individuals.

Diseases typical of late adulthood, notably joint disease, were identified in a considerable number of individuals. Osteoarthritis was also common, particularly in the hands and wrists, and also in the knees in females and the hips in males. Ten individuals, mostly women, apparently had osteoporosis, and an 82-year-old man suffered from Paget's disease. Diffuse idiopathic skeletal hyperostosis (DISH), a disease often linked with rich diet and obesity, was found in nine males and a female, all in the older age category. Interestingly, although syphilis was supposedly rife during this time no evidence was found for this.

Congenital malformations were rare. Possible indicators of childhood illness or malnutrition were found in several subadults and adults. Evidence for active rickets was found in one child aged 1–5 years and healed rickets in an 11-year-old. Fractures mainly affected men and were predominantly in the upper body; they were generally well aligned and well healed with no indication of secondary infection. Post-mortem examinations had been undertaken on two males.

ACKNOWLEDGEMENTS

The archaeological fieldwork, post-excavation assessment and publication were generously funded by the London Diocesan Fund and F L Estates (Old Church Street) Ltd. The Museum of London Archaeology Service (MoLAS) and the Parochial Church Council of Chelsea Old Church are most grateful to the Cadogan Estate, which made a substantial contribution towards the cost of the project.

MoLAS is especially grateful to the Reverend Peter Elvy, vicar of Chelsea Old Church at the time of the excavation, and the staff of the Church, for their cooperation and considerable assistance. Particular thanks go to Angus Stephenson and Duncan Hawkins of CgMs Consulting, who commissioned and monitored the archaeological work on behalf of F L Estates (Old Church Street) Ltd, and to John Simpson (the architect), Simon Blausten of Cyril Leonard & Co, Simon Coe and Bob Wilson of Biscoe Craig Hall. The fieldwork was monitored for English Heritage (Greater London Archaeology Advisory Service) by Catherine Cavanagh. Thanks also go to the staff of Necropolis who worked closely with MoLAS throughout the excavation.

The fieldwork was supervised by Robert Cowie. The excavators were Naomie Arazi, Nathalie Cohen, Neville Constantine, Vicki Ewens, Nicola McKenzie, Adrian Miles, Ceri Rutter, David Thorpe, Aidan Turner, Keith Webster and Robin Wroe-Brown. Jessica Cowley and David Mackie were responsible for surveying, and Maggie Cox and various site staff for on-site photography.

Contributions to the post-excavation assessment were made by the following: Philippa Bradley (flint), Brian Connell and Bill White (human bone), Robert Cowie (site records and genealogical research), Rupert Featherby (Roman pottery), Kieron Heard (clay tobacco pipes), Richard Hewett (genealogical research), Jackie Keily (accessioned finds), Adrian Miles and Valerie Griggs (coffin furniture), Jacqueline Pearce (medieval and post-medieval pottery), Alan Pipe (animal bone), Susan Pringle (building material) and Kate Pollard and Ant Sibthorpe (digitisation of the site plans).

The authors would like to thank Hazel Cook and David Walker, Royal Borough of Kensington and Chelsea Libraries and Arts Service, for assistance with documentary research, and Robin Darwall-Smith for information concerning the Butler family. Robert Cowie is grateful to Josephine Brown of Pre-Construct Archaeology Ltd for providing plans of the excavation at 6–16 Old Church Street, and to Lesley Hannigan for transcribing wills. Jelena Bekvalac and Tania Kausmally are grateful to Rachel Ives for information concerning osteoporosis and to Brian Connell for his advice and assistance.

The summary was translated into French by Elisabeth Lorans and into German by Manuela Struck. The index was compiled by Susanne Atkin.

1

Introduction

1.1 Location and circumstances of fieldwork

This report provides an account of the archaeological evaluation and excavation undertaken on the site of 2–4 Old Church Street, Chelsea, by MoLAS. The site was on the east side of Old Church Street immediately to the north of All Saints, Chelsea Old Church. The excavation area included the sites of an 18th- and 19th-century school and vestry (Petyt House), a 1960s vicarage (4 Old Church Street) and church hall (also called Petyt House) and the northern part of the churchyard. The Ordnance Survey National Grid reference for the centre of the site is 527087 177617 (Fig 1).

The church, which has existed since the 12th century, lies at the historic heart of Chelsea in an area now recognised as an Archaeological Priority Zone by the local authority. It is perhaps most famously linked with Sir Thomas More, who lived in Chelsea from 1524 to 1534 and regularly worshipped there. Many histories also recount how Henry VIII secretly married Jane Seymour there, although this is almost certainly apocryphal.

On the night of 16–17 April 1941 several buildings at the south end of Old Church Street, including Petyt House, were destroyed during a bombing raid (Matthews and Bell 1957, 7–26; Russett and Pocock 2004, 130–3). Most of the church was also destroyed, although the south chapel (named after Thomas More who rebuilt it in 1528 as his private chapel) was relatively unscathed. Monuments in the churchyard, which was buried under rubble from the surrounding buildings, would probably also have suffered serious damage.

The church was rebuilt during the 1950s as an exact copy of its precursor. The new building, which was reconsecrated in 1958, incorporated the surviving south chapel and a number of monuments salvaged from the ruins of the bombed church. By the early 1960s a new vicarage and church hall (Petyt House) had also been built. The latter was considerably larger than its earlier namesake and was set further back from the road, so that it extended across much of the northern part of the churchyard, which had been used as a cemetery during the 18th and first half of the 19th century. However, prior to the construction of the church hall permission had been granted by Parliament for the clearance of burials (All Saints Chelsea Act 1959).

In 2000 the vicarage and Petyt House were demolished to make way for a new building in Georgian style designed by the architect John Simpson. Archaeological fieldwork followed soon after demolition as a condition of planning consent for the redevelopment of the site. It began in April 2000 with the excavation of a single evaluation trench. This exposed features associated with post-medieval and earlier settlement. It also revealed the north wall of the churchyard wall and up to fifteen 18th- and 19th-century burials to the south. The discovery of the burials was unexpected, as it had been assumed that all human remains had been removed from the site prior to the construction of the church hall some 40 years earlier.

The main excavation followed in two stages later in the year.

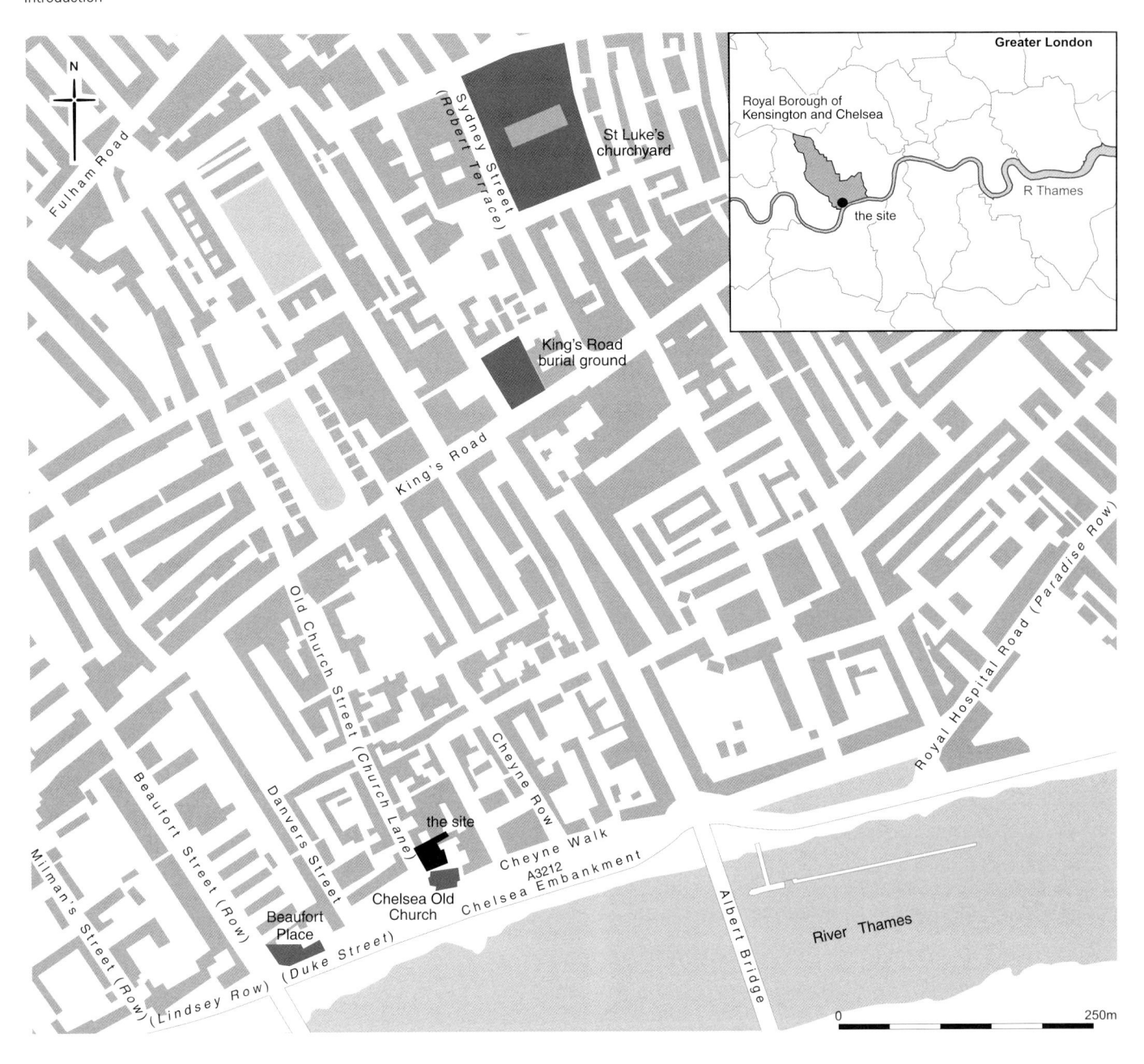

Fig 1 *Location of the site and nearby parish churchyards and streets mentioned in the text (former street names in italics) (scale 1:6000)*

The first phase was from May to August, while the second was from October to November. During this work the entire area of the proposed new building was investigated (Fig 2; Fig 3). By the end of the excavation evidence for prehistoric activity and Roman, medieval and post-medieval settlement had been recovered together with the remains of 290 parishioners buried in the 18th and first half of the 19th century (Cowie 2002).

1.2 Organisation of this report

The report is divided into three principal parts. The archaeological sequence concerned with local topography and settlement is described in chronological order in Chapter 2.

This is followed, in Chapter 3, by a description and discussion of evidence recovered from the churchyard relating to various aspects of burial practice and the personal and family histories of identified excavated individuals. Chapter 4 considers the burial population with particular reference to information obtained from the examination of skeletal remains and comparative data from other post-medieval cemeteries.

Historical and genealogical evidence is briefly outlined in Chapters 2, 3 and 4 in order to place the archaeological finds in context. However, for further information about Chelsea Old Church and its environs the reader is directed to local histories by Faulkner (1829), Rev L'Estange (1880), Beaver (1892), Blunt (1921), Bowack (1705), Rev Stewart (1932) and Denny (1996). *Survey of London* volumes 2, 4 and 7 (Survey of London 1909; 1913; 1921) and the recently published *Victoria County History* on Chelsea (*VCH* 2004) provide authoritative accounts.

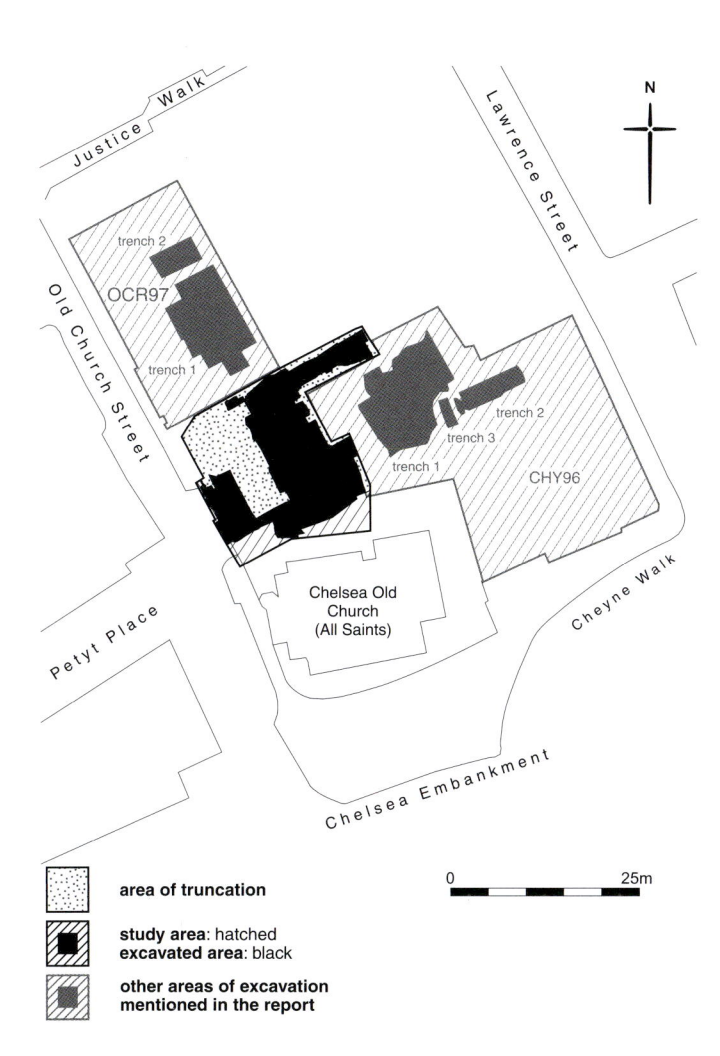

area of truncation

study area: hatched
excavated area: black

other areas of excavation
mentioned in the report

0   25m

The structure of the pre-war church is described in detail by Walter Godfrey (Survey of London 1921) and is summarised by RCHME (1925, 7–11). Russett and Pocock (2004) give further background information on the development that led to the excavation.

All finds and environmental samples were recorded to current MoLAS standards and entered on to the Oracle database, and full reports have been deposited as part of the site archive. The archives are held under the site code OCU00 in the London Archaeological Archive and Research Centre (LAARC) at Mortimer Wheeler House, 46 Eagle Wharf Road, London N1 7ED, where they may be consulted by prior arrangement with the Archive Manager.

1.3 Textual and graphical conventions

The basic unit of reference in the site archive and this report is the context number. This is a unique number given to each archaeological feature or stratum representing a single action (such as a layer, wall, grave cut etc). Context numbers appear in

Fig 2 *Location of the archaeological investigations and the earlier adjacent investigations at 6–16 Old Church Street (OCR97) and 61–62 Cheyne Walk (CHY96) (scale 1:1000)*

Fig 3 *Excavation site viewed from the church tower, looking north*

the text in square brackets, for example [10]. Museum of London accession numbers given to certain artefacts from the site are shown thus: <10>. The illustrated pottery has been assigned a catalogue number and is shown thus: <P1>.

The archaeological sequence is described in terms of land use and is divided chronologically into six periods (1–6) which are based on a combination of artefactual dating, stratigraphic development and documentary sources. The land-use elements comprise Building (B) and Open Area (OA).

Adjacent archaeological excavation sites at 6–16 Old Church Street and 61–62 Cheyne Walk are referred to by their respective site codes, OCR97 and CHY96 (Fig 2, Partridge 1997; Farid 2000). As far as possible information about the human remains recovered from the churchyard is presented in non-technical language with explanation of specialist terms.

All the ceramic tile fabric types mentioned in the text are represented in the fabric reference collection which is available for consultation in the London Archaeological Archive and Research Centre (LAARC) on request. Descriptions of each fabric type discussed in the text are also available from the LAARC.

The clay tobacco pipes are dated and classified according to the 'Chronology of bowl types' (Atkinson and Oswald 1969, 171–227) or the 'Simplified general typology' (Oswald 1975, 37–41). The prefixes AO and OS are used below to indicate which typology has been applied.

The graphical conventions used in the plans in this report are shown in Fig 4.

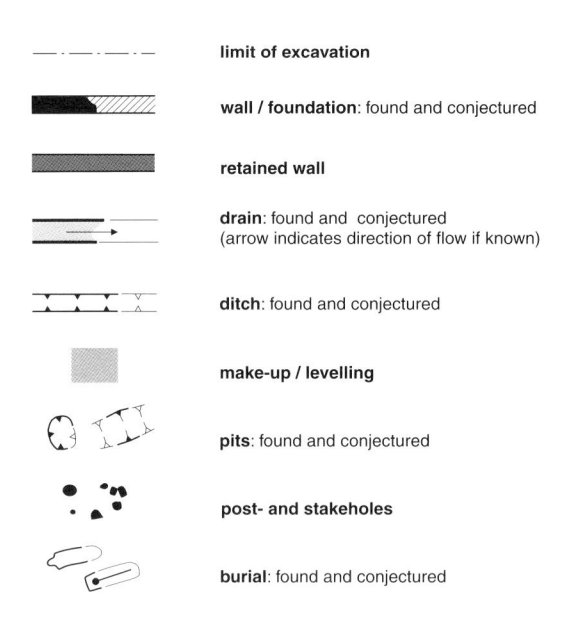

——·——·——	**limit of excavation**
	wall / foundation: found and conjectured
	retained wall
	drain: found and conjectured (arrow indicates direction of flow if known)
	ditch: found and conjectured
	make-up / levelling
	pits: found and conjectured
	post- and stakeholes
	burial: found and conjectured

Fig 4 Graphical conventions used in this report

2

The archaeological evidence

2.1 Geology and topography (period 1)

The natural landscape

The site is located within the historic core of Chelsea, which occupies a fairly level, low-lying area on the Middlesex bank of the Thames. The Thames was of particular importance to the development, economy and character of Chelsea until the early 1870s, when the construction of Chelsea Embankment cut a swathe between the settlement and its waterfront. At one time two local tributaries of the Thames, the Westbourne and Counter's Creek, respectively marked the eastern and western boundaries of the parish (Barton 1992, 43–7; *VCH* 2004, 1).

The geology (OA1)

The site lies on drift deposits of the first River Terrace (British Geological Survey 1981), which locally comprise light yellow-brown to orange-brown sand with occasional thin lenses of gravel. The surface of the terrace deposits sloped gently down towards the river, from about 4.90m OD on the north side of the site to 4.30m OD in the south-west.

2.2 Prehistoric (period 2)

Archaeological background

Relatively few prehistoric artefacts have been found in Chelsea, although they include a small number of residual struck flints and potsherds recovered during excavations next to the site (CHY96 and OCR97; Partridge 1997; Farid 2000). By contrast, the adjacent stretch of the Thames, Chelsea Reach, has long been a rich source of prehistoric artefacts, especially Bronze Age and Iron Age weapons (Lawrence 1929, 92–3). Recent finds from the river include a Neolithic wooden club recovered from a peat bed at Chelsea Harbour (Webber 2004). Some objects may have been discarded or lost accidentally, but it is thought that many were votive offerings (Bradley 1990). Such ritually deposited items may include the famous Battersea shield, which was probably found during the construction of Chelsea Bridge in 1851–8, as were numerous human skulls (Cuming 1857, 238; Cotton 1999). Two skulls from the river at Battersea and between Battersea and Vauxhall Bridges were respectively dated by radiocarbon assay to the Neolithic period and the Bronze Age (Bradley and Gordon 1988, 508). A large fragment of a trepanned skull recently found on the foreshore at World's End, Chelsea, has also been dated to the Bronze Age (Mays and Sidell 2003).

Pits (OA2)

Among the earliest features on the site were two pits [683] and [817], which may have been prehistoric in date (Fig 5). They

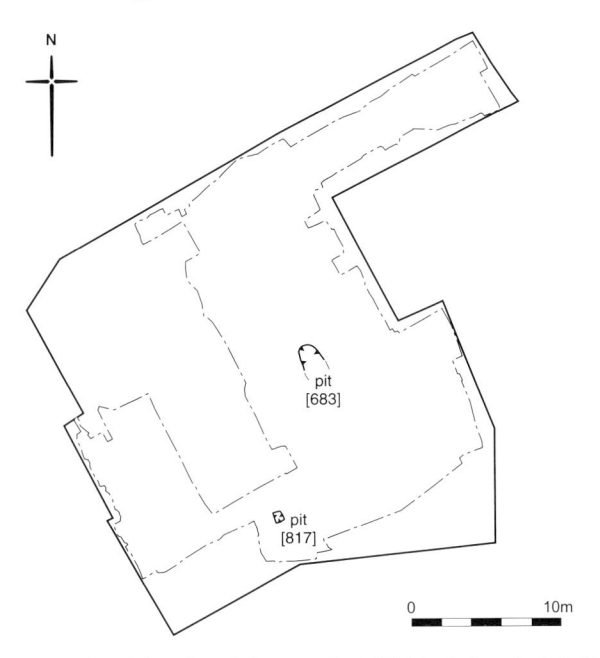

Fig 5 *Plan of the prehistoric features at 2–4 Old Church Street (scale 1:500)*

respectively produced a mesolithic burin <80> and a small undiagnostic flake.

Up to eight other pieces of worked flint were recovered from the site either as unstratified finds or from features mainly of Roman and medieval date. Generally they comprise flakes of good quality, brown to black flint with a thin grey or buff cortex, one of which has been slightly retouched. One piece, from a slot of medieval or later date, might be Roman or later building material, but the other flints are probably prehistoric.

2.3 Roman (period 3)

Archaeological background

In the Roman period the site lay about 6km upstream from Londinium and 2km south of the conjectured route of a major road roughly on the line of Kensington Road, which emerged from the provincial capital at Ludgate (to the north-east) and apparently joined the main Silchester Road at Chiswick (to the west) (Merrifield 1983, 121–2).

Previous excavations at OCR97 had revealed a ditch and pit which were thought to represent a small farm. Roman finds from this site comprised 3rd-century AD pottery, tile fragments dated to c AD 55/70–140/200, nails and basalt lava querns (Farid 2000, 120, 125–6). The only other finds of Roman date from the immediate area were a few residual potsherds from CHY96 (Partridge 1997) and eight stamped pewter ingots from the Thames near Battersea Bridge (RCHME 1928, 175).

A timber building (B1)

Two parallel slots near the north side of the site might indicate

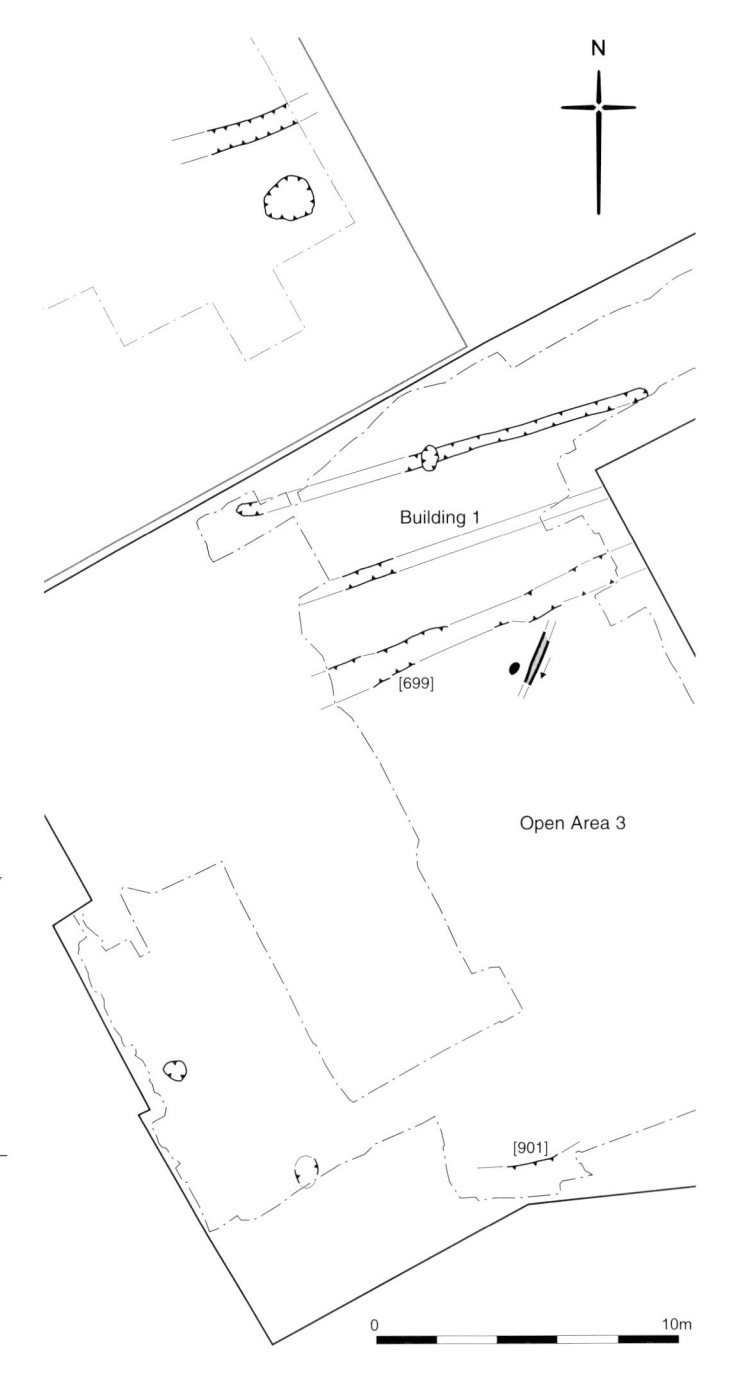

Fig 6 *Plan of Roman features at 2–4 Old Church Street in relation to those at a neighbouring excavation site (OCR97) (scale 1:250)*

the presence of a simple rectangular timber building (Fig 6). Although they did not produce any artefacts both were early in the site sequence and the northern slot was cut by a Roman pit. The putative building would have been roughly parallel to the river, and was on similar alignment to ditches dated to the Roman period.

Ditches and pits (OA3)

There were two ditches in Open Area 3 (Fig 6). One ditch, [901], lay on the south side of the site and survived to a depth of 0.39m. The ditch may have been relatively early, for it contained five sherds of an organic-tempered pottery dated to the Late Iron Age or early Roman period and therefore broadly contemporary with residual pottery found at CHY96. It also produced fragments of Roman roof tile of a type (fabric group 2815) manufactured between *c* AD 50 and 160. The other ditch, [699], lay about 2m to the south of Building 1. It was up to 1.1m wide and survived to a depth of 0.30m. Pottery dated to *c* AD 120–250 and a residual flint flake were recovered from its fill.

Other features dated to period 3 comprised a gully, a posthole and three pits, which survived to depths of between 60mm and 0.28m. All contained Roman pottery. In addition, one pit produced a fragment of Roman brick in fabric group 2815.

Pottery

A total of 53 sherds of Roman pottery were found. The pottery was recovered mainly from the pits although a small quantity came from the ditches, and some was found in later features. Apart from the pottery from the southern ditch most of the assemblage appears to date to the late Roman period, and is similar in composition and date to that found at OCR97. Pottery identified as Alice Holt/Surrey ware was probably from later early to mid 3rd-century AD phases of the industry rather than the earlier 1st to 2nd-century AD phases.

Building material

Further evidence for Roman occupation was provided by 33 pieces of Roman ceramic building material. These were found mainly as residual fragments in medieval pits, but pieces of tegula, imbrex and brick, and a single tessera were recovered from a Roman pit and ditch (OA3). Most are in the red-firing clays (fabric group 2815) typical of London production *c* AD 50–160, although one imbrex fragment is in a later variant of the local fabric produced between *c* AD 140 and 250 (fabric 2459B). One piece of brick is in a non-local fabric (3009) with a date range of *c* AD 100–20, which may have been made in Hampshire.

Discussion: Roman activity

The Roman features in the excavation area and at OCR97 probably represent the remains of a rural settlement with associated field and enclosure ditches. The pottery from both sites suggests that the main phase of occupation was in the 3rd century AD. However, pottery from the southern ditch and residual pottery from CHY96 suggests that some activity dated to the 1st century AD. The lack of evidence for Roman activity at other nearby excavation sites suggests that the settlement occupied a fairly small area, and was probably an isolated farm.

The alignment of the building and the ditches may have been determined by local topography for they were roughly parallel to the Thames. The other main axis of the putative field system would probably have been perpendicular to the river – assuming that the fields and enclosures were roughly rectangular and coaxial. The most direct route between the settlement and the main road to the north would also have been on this alignment, and would have avoided the extensive tracts of fen that lay to the east in what is now Pimlico and Westminster. This raises the possibility that Old Church Street may have begun as a farm track in the Roman period rather than as a Saxon or medieval road.

Evidence for rural settlements of this period is extremely rare in central London, and the study of such sites continues to be a high priority (Perring and Brigham 2000, 156). Why so few rural settlement sites have been found close to Londinium is not fully understood, although a lack of archaeological fieldwork is probably partly to blame. Another possible explanation is that the surrounding land may have been mainly cultivated by people living in the town (ibid, 153). Thus the decline of the urban population in the late 2nd and 3rd centuries AD may have allowed the settlement at Chelsea to develop. However, it has been argued also that a corresponding reduction in demand for foodstuffs could have had the opposite effect on some farms in Londinium's hinterland (Howe 1998, 29).

2.4 Saxon and medieval (period 4)

Archaeological and historical background

Chelsea (Caelichyth, Celchyth) is first mentioned in charters issued at ten Church councils held there between AD 785 and 816 (Gover et al 1942, 85–6; Cubitt 1995, 28–9, 308–9). The frequency of the councils, some of which were attended by the Mercian kings Offa and Coenwulf, suggests the possible presence of a Saxon minster and/or royal estate centre. The later village lay 2km south of the important Saxon land route known by the 11th century as Akemannestraete; it ran along the line of the former Roman road. Some form of Middle Saxon settlement might be indicated by evidence for a post-built structure and a pit dated to *c* AD 650–750 at OCR97 (Farid 2000, 120–1), and two fish traps with calibrated radiocarbon dates of AD 660–900 and AD 650–890 on the foreshore at Chelsea Harbour (Cohen in prep).

Domesday Book records that the manor at Chelsea was held by Wulfwen during Edward the Confessor's reign (1042–66), although there is no clear archaeological evidence for Late Saxon settlement in the area. At the time of the survey in 1086 the manor comprised two hides with enough arable land to support five plough teams, suggesting a relatively small rural settlement (Morris 1975, 20.1).

The medieval village extended along the riverbank on either side of the church and the adjacent manor house. Its two principal streets comprised a riverside road and Church Lane (later Old Church Street).

The manor house and its courtyard, ancillary buildings, gardens and a 4-acre close lay on the north and east sides of the church (*VCH* 2004, 113). Medieval pits, ditches and other features were recorded in this area during excavations at CHY96 and OCR97 (Fig 7, Partridge 1997; Farid 2000, 121–3). The features at CHY96 also included three parallel gullies or bedding trenches, which may have been dug for the cultivation of vines. Most of the pottery from this site was dated to *c* 1080–1200, while the assemblage from OCR97 was more broadly dated to *c* 1050–1350. The main entrance to the manorial complex may have been in Church Lane, and possibly survived as a passage shown on a plan of 1706 crossing the northern part of the site between Church Lane and Lawrence Street (*VCH* 2004, 113).

It is not known exactly when Chelsea gained its church. There is no indication for its existence in Domesday Book, and it is first mentioned in 1157, when Pope Adrian IV confirmed a grant of the church to the abbey of Westminster (Beaver 1892, 8; Stewart 1932, 17). Subsequent references are made to a rector of Chelsea in 1230 and again to the church in 1290. Before wartime bombing the earliest surviving parts of the church were the eastern end of the chancel and the north chapel, respectively dated to the 13th and 14th centuries (Survey of London 1921, 1, pl 15; RCHME 1925, 7–8).

Medieval soil, pits and a ditch (OA4)

A soil horizon comprising yellow-brown sandy silt formed on the site during this period (layer [2], Fig 7). It apparently sealed some medieval pits and was cut by others. The soil produced a body sherd from a cast copper-alloy vessel, a fragmented piece of sheet copper alloy with the remains of rivets or decoration, several fragments of peg tile and 27 sherds of pottery (429g). Three of the latter date to before *c* 1150/ 1180, but most were probably deposited between *c* 1300 and *c* 1450 (the latest date to after *c* 1430); four post-medieval sherds are presumed intrusive.

Over two dozen medieval pits were found. Most dated from the second quarter of the 13th century to the mid 14th century. They were mainly concentrated in the north-east quarter of the site, where there had been less disturbance by later activity, but four survived near the south-west corner of the site.

A row of five closely spaced pits on the north side of the site (Fig 7; Fig 8) was aligned roughly perpendicular to the medieval road (Old Church Street) and may have run along one side of a property boundary. The pits were between 1.4m and 1.5m in diameter and 0.35–0.85m deep, and were filled mostly with brown sandy silt with a distinctive reddish hue quite unlike the fills of most of the other pits. Four of the pits each contained one or two sherds of pottery dated to *c* 1230–1350/ 1500; the latest diagnostic find is a sherd of Tudor Green ware (*c* 1350–1500). The pits also contained pieces of peg tile, some with patches of glaze, a characteristic of some roofing tile produced during the 12th to 14th century. One pit produced the remains of an iron rowel spur <120>, which probably dates to the 14th century and appears to have been tinned to give it a silver appearance (cf Ellis 1995, 140, fig 99, nos 338–9). Another produced the neck of a Kingston-type ware roof finial dating to *c* 1240–1400.

The other pits were apparently randomly scattered. They survived to depths of between 0.12m and 0.65m and yielded pottery and/or fragments of peg tile. Fifteen contained pottery (mostly less than 12 sherds in each) of similar date. The largest group is from pit [325], which contained 28 sherds (435g) dating to *c* 1270–1350. One pit [671] also produced the remains of an iron pintle or hinge pivot <121>.

The only other medieval feature was a ditch [3], which was aligned roughly parallel to the medieval road and at right angles to the row of pits. It was up to 0.63m wide and survived to a depth of 0.15m. Three small sherds of pottery dated to *c* 1080– 1350 and fragments of peg tile were recovered from its fill.

Pottery

No Saxon pottery was found, and the ceramic evidence for any activity before *c* 1200 is limited to 11 sherds. Most of the 134 sherds (1697g) of medieval pottery comprise domestic material dating to *c* 1230–1350. The fabrics mainly comprise London-type wares (37 sherds, 466g), coarse Surrey-Hampshire border ware (25 sherds, 470g) and Kingston-type ware (24 sherds, 213g), with 20 sherds of south Hertfordshire-type greyware, seven of Mill Green ware and one of Earlswood-type ware. Only one sherd from a Cheam whiteware barrel jug can definitely be dated to the 15th century. Small abraded sherds of residual medieval pottery, including a sherd of Saintonge polychrome ware, also occur in periods 5 and 6.

Building material

A few fragments of medieval building material, found in later features, probably came from either the church or a high-status building in the locality. They include three pieces of two-colour decorated floor tiles (fabrics 1810, 1811) of the type made in Penn, Buckinghamshire, between *c* 1350 and 1390. All are of the same design (E2230/P52; Hohler 1942; Eames 1980) and are likely to have come from a single later 14th-century floor. A block of moulded Reigate stone <169> from an arched feature, probably a doorway, appears to be of similar date (T P Smith, pers comm). The moulding has traces of graffiti and is thus likely to have come from a fairly low level in the structure. Another fragmentary tile in fabric 2322 has plain green glaze over white slip and is dated to the 14th to late 15th century.

Discussion: Saxon and medieval activity

The medieval features in the excavation area and on the adjacent sites of CHY96 and OCR97 lay within the grounds of the manor house, which lay to the north and east of the church. They were probably associated with seigneurial gardens, which may have been used for leisure, food production and possibly viticulture (Dyer 1994, 114–16). Most of the pits were probably used for rubbish disposal, although the row of pits on the north side of the site was probably dug for planting trees or shrubs. The latter may have lined the possible entranceway leading from Church Lane.

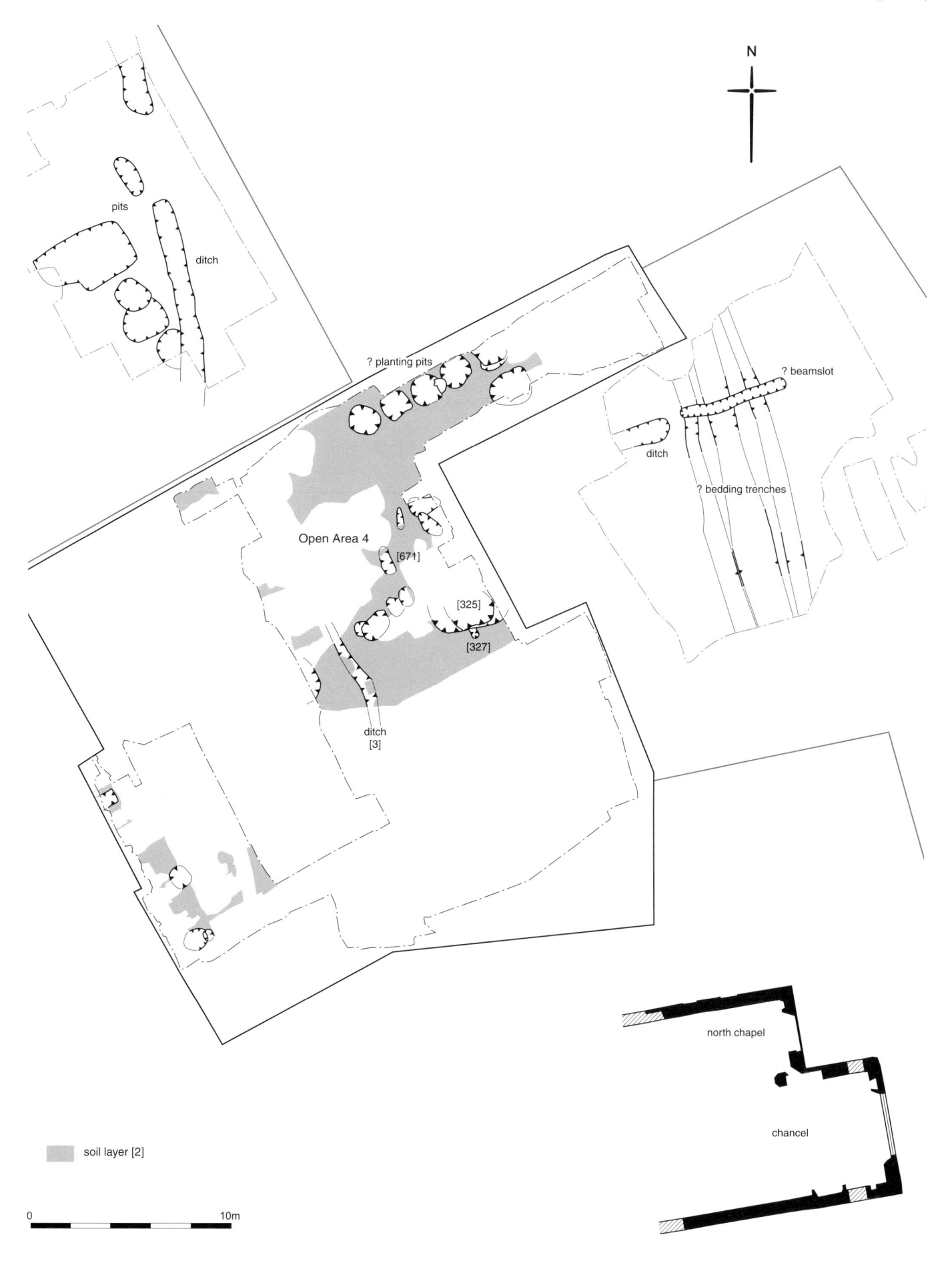

N

pits

ditch

? planting pits

? beamslot

ditch

? bedding trenches

Open Area 4

[671]

[325]

[327]

ditch
[3]

north chapel

chancel

soil layer [2]

0   10m

Fig 7 Plan of medieval features at 2–4 Old Church Street in relation to those at neighbouring excavation sites and to the medieval church (scale 1:250)

Fig 8 Row of medieval pits on the north side of the site, looking south-west

As at OCR97 (Farid 2000, 121; Jarrett 2000a, 138–9) there is no definite ceramic evidence of any activity before the probable foundation date of the church in the early to mid 12th century. It is possible that the centre of Saxo-Norman/early medieval activity was closer to CHY96, where most of the medieval pottery dates to *c* 1080–1200 (Blackmore 1997; Farid 2000, 123).

There is little evidence on any of the three sites for medieval activity after *c* 1350/1400. This suggests that the population of the settlement may have declined in the mid to late 14th century, possibly as a consequence of the arrival of the Black Death in London in 1348–9.

2.5 Chelsea in the 16th and 17th centuries (period 5)

The village

During this period Chelsea was popular with courtiers and royal officials, who were attracted by its tranquil riverside location some distance from the main highway but a conveniently short journey by boat to Westminster and Whitehall (*VCH* 2004, 18–31). It also provided an occasional refuge from the plague during its visitations on London. Consequently the settlement had a number of great Tudor houses, notably Henry VIII's manor house, granted him by Lord Sandys in 1536, Shrewsbury House and Thomas More's house, which was later extended and renamed Beaufort House (Survey of London 1913, 18–27; *VCH* 2004, 18–22). Other mansions were built in the 17th century, including Danvers House. Today, all that survives of these great houses are a number of 16th-century garden walls (Survey of London 1913, 61; RCHME 1925, 13).

Buildings of lesser status included taverns and dwellings for farmers, artisans and watermen. There was also a school, which was built in 1603 by Richard Ward, rector (1585–1615), on waste ground next to the church (Faulkner 1829, ii, 144; Davies 1904, 188–9; Survey of London 1913, 53; *VCH* 2004, 176, 183). Its site (later occupied by Petyt House) apparently belonged to the parish although consent to build was given by the lord of the manor, the Earl of Nottingham.

Chelsea did not change significantly in size until the second half of the 17th century, when the pace of growth quickened considerably, especially after the construction of the Royal Hospital in 1682–92 (*VCH* 2004, 26). One area particularly affected by building work at this time was the 4-acre site of the old manor house and its grounds immediately to the north and east of the church. In 1687 the estate, then owned by the Lawrence family, was leased to Cadogan Thomas of Southwark for redevelopment (ibid, 31). The old manor house was probably demolished at an early stage in this work, and by the time of Thomas's death in 1689 several substantial brick houses had been built in Church Lane, Lawrence Street and Johns Street (probably Justice Walk) (Fig 2).

The Old Church was added to and modified several times during this period (Fig 9). Notable works included the rebuilding of the south chapel in 1528 by Sir Thomas More, and various modifications in the late 17th century. The latter included the enlargement of the nave and reconstruction of the church tower in 1669–74 to accommodate the growing congregation (Faulkner 1829, i, 201; Survey of London 1921, 3). As part of the late 17th-century improvements the churchyard was raised considerably and enclosed with a high brick wall. (Probably the wall on the north side of the church not visible in Fig 9.)

A cellar of a house in Church Lane (B2)

The remains of a cellar (B2, Fig 10) lay on the north side of the churchyard, immediately adjacent to Old Church Street. They consisted of a wall and part of a barrel vault built of soft orange-red brick (fabric 3039) of a type made between *c* 1450 and *c* 1700 and usually used in high-status buildings. The floor of the cellar was tiled. The cellar probably belonged to a small rectangular building shown roughly in this location on James Hamilton's plan of 1664 (Fig 11) and on a sketch plan of 1706 by an anonymous draughtsman.

Fig 9   View of Chelsea Old Church from the south-east (c 1750), an engraving by J Roberts of a picture by Jean Baptiste Claude Chatelain (Guildhall Library, City of London)

Yards and gardens in Church Lane: pits (OA5)

A few features are dated to the 16th century and a somewhat larger number to the 17th century. Those in the northern half of the site (OA5, Fig 10) lay outside the old churchyard in the back gardens or yards of houses in Church Lane (now Old Church Street). They mainly comprised rubbish pits and quarries, but also included a brick-lined cesspit and a well.

Pits [335] and [415], and [1160] (OA6) may have been the earliest features dated to this period, for each contained one or two sherds dating to c 1480–1550. Pits [128] and [290] contained only redwares, dating to c 1480–1600 and c 1580–1650 respectively, but most other pits contained redwares and/or Surrey-Hampshire border ware. Pits [75] and [56], dated to c 1550–1650 and c 1575–1700 respectively, also contained Frechen stoneware. Pits [308] (OA6) and [377] contained tin-glazed ware and date to after c 1630. The latter also produced a Nuremburg jetton <140> of Hans Krauwinckel II (1583–1635) in very good condition. These coin-like objects or reckoning counters were used in the calculation of accounts. Nuremburg was the prime source of jettons used in Britain from the 16th century and they are a common find from this period. Bottle and/or window glass was recovered from pit [308] dated to the late 17th century.

Waste ground: quarries and pits (OA6)

Quarry pits

During the 17th century an area of waste ground to the north of the church, which later became the north-east corner of the churchyard, was stripped of soil exposing the underlying terrace sand, into which several deep quarry pits were dug (OA6, Fig 10). Three of these contained small groups of later 16th- and 17th-century pottery including Martincamp and Frechen stonewares and the bases of two birdpots (Blackmore 2004, 276–7), as well as fragments of pantile, which first came into use in London in the 1630s. Four other pits yielded sherds of tin-glazed ware with 'Chinaman in grasses' and Persian blue decoration, dated to c 1670–1700. Two of these also produced fragments of clay tobacco pipes broadly dated to c 1680–1710, and part of the base of a 17th-century beaker <117> in colourless glass with white trail decoration. The latter may have been a *façon de Venise* vessel made in England or an Italian import.

The evidence suggests that the quarrying took place in the mid to late 17th century. Sand from the pits was probably used in the rebuilding of the church nave and tower or possibly for other nearby construction projects, such as the extensive building work on the Lawrence Estate, which began in the late 1680s.

11

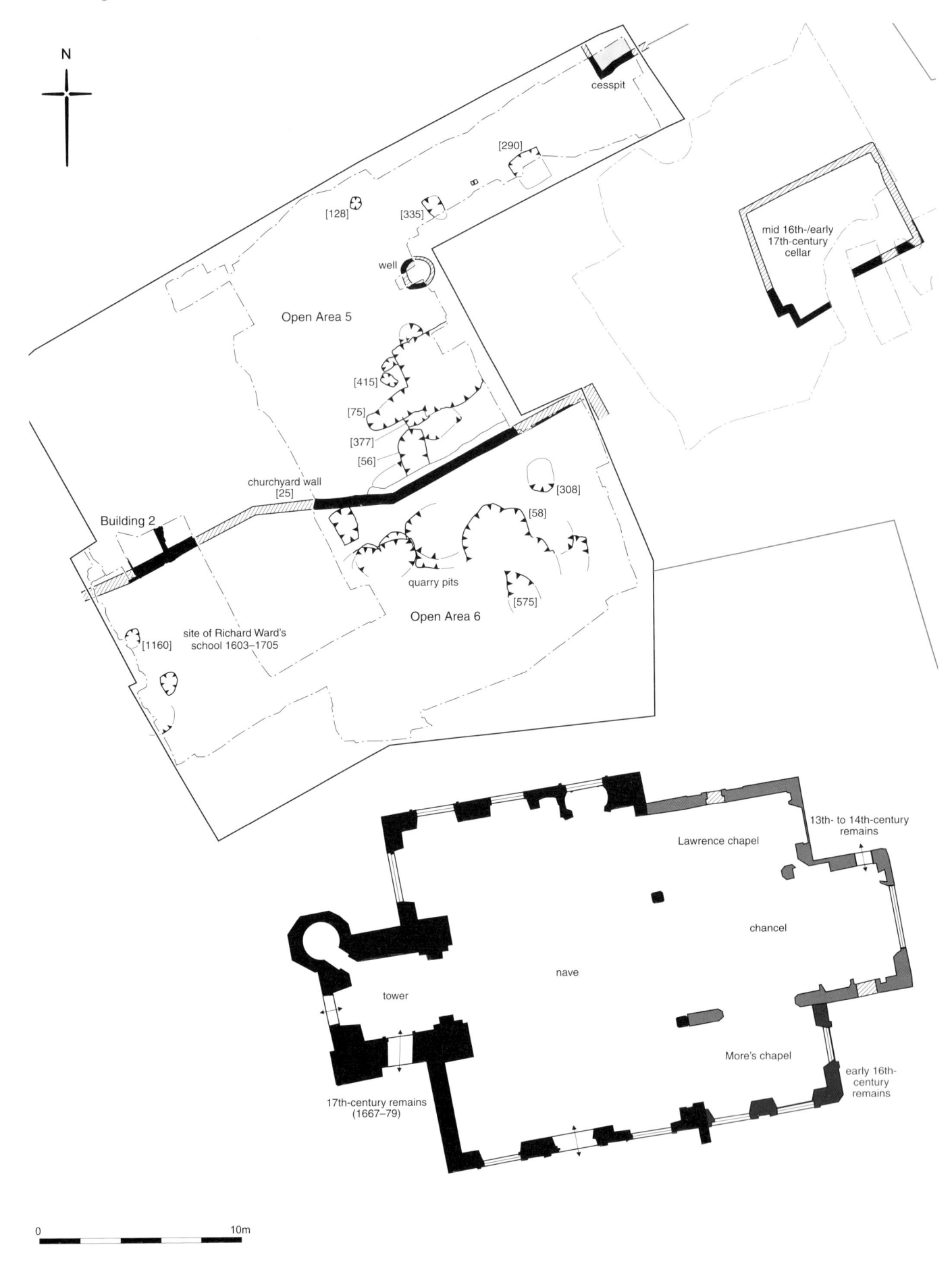

Fig 10 Plan of 16th- and 17th-century features at 2–4 Old Church Street, neighbouring excavation site CHY96 and the church (scale 1:250)

The finds indicate that the quarry pits could not have been filled before about 1680 at the earliest. The pits, however, must have been filled by the mid 1690s or the first decade of the 18th century, by which time the area had begun to be used as a cemetery. The absence of residual human bones in the quarries suggests that there had been no previous interments in the area.

Pits

Four or five small pits cut surviving areas of topsoil in the western part of the churchyard (Fig 10). These may have been associated with nearby domestic occupation, and were possibly associated with the parish clerk's house in the north-west corner of the churchyard.

The churchyard wall

A wall marking the northern limit of the churchyard ([25], Fig 10) was made of red brick (fabric 2271) of a type manufactured between c 1450 and c 1700. Documentary sources suggest that its construction was part of a programme of major improvements to the church that began in 1669. However, it is unlikely that the wall was built before the quarries (OA6) were filled, so that a construction date after the 1680s or 1690s seems more probable. After the wall had been built a layer of earth was dumped over the newly enclosed churchyard to raise and level the ground. The earth lay against the well-pointed southern face of the wall.

Pottery

The post-medieval pottery amounts to 149 sherds (3845g). There are few imports, and the 16th- to 17th-century forms mainly comprise standard domestic material, but the redwares also include a sprinkler pot (pit [290]), two birdpots (pit [58] and quarry pit [575] in the churchyard) and vessels possibly used in dairying.

Building material

Early post-medieval building materials from later features in Open Area 6–Open Area 8 provide evidence for high-status buildings in the locality. Notable finds include part of a stove tile (fabric ?2310) dated to between c 1470 and c 1650 and two fragments of polychrome tin-glazed floor tiles (fabric 3067) dated to the late 16th or early 17th century. The stove tile has green glaze over a white slip, and is decorated with a possible foliate design. Both floor tiles have strapwork designs on white backgrounds; one in yellow ochre, the other in blue, green, yellow and orange. Other finds include a fragmentary Flemish floor tile (fabric 2850) and a piece of floor tile (fabric 3080), both with clear glazes respectively in olive-brown and brown and dated to between the late 15th and 17th centuries.

An ivory comb

Among the few unstratified finds from the site is a double-sided ivory comb <234> with fine and coarse teeth. These are characteristic of the 16th and 17th centuries and are often depicted in 'genre' paintings of the period, being used to remove lice from the hair (Margeson 1993, 66).

2.6 Chelsea in the 18th and 19th centuries (period 6)

From riverside resort to London suburb

This period saw the gradual transition of Chelsea from a large riverside village to a London suburb (*VCH* 2004, 26–31). At the beginning of the 18th century Chelsea was a substantial settlement of about 300 houses, although it maintained a rural aspect and was surrounded by fields, orchards, nurseries and market gardens (Fig 11). Indeed, until the early years of the 19th century many of its inhabitants continued to work the land. By the middle of the 18th century Chelsea had become a fashionable resort for Londoners, especially after the opening of Ranelagh Gardens in 1742. Descriptions of Chelsea at this time suggest that it was a relatively healthy and prosperous place compared with many parts of nearby London. Nevertheless, there were poor in the parish, and in 1737 a workhouse was established in Chelsea for their employment (Faulkner 1829, ii, 24–5).

During the first half of the 19th century Chelsea became increasingly urban in character as the surrounding fields rapidly disappeared beneath new streets and houses. In 1819–24 a new parish church, St Luke's, was built in Sydney Street to serve a settlement that by then sprawled well to the north of King's Road (Fig 1). Its predecessor in Church Street became a chapel of ease, but when the old parish was revived in 1951 it returned to use as a parish church, becoming All Saints, Chelsea Old Church (Russett and Pocock 2004, 113, 138–9).

A cellar at the back of a house in Old Church Street (B3)

The east end of a cellar (B3) was partially exposed near the north side of the site (Fig 12). Its walls were made of unfrogged red bricks bonded with whitish grey lime mortar. It would have been located at the back of the terrace house (no. 4), shown on 18th- and 19th-century maps (eg Fig 11).

Wells and pits (OA8)

In the gardens and yards to the north of the churchyard there were a small number of features generally indicative of domestic occupation and gardening (Fig 12), although the presence of crucibles in an 18th-century pit [61] also suggests

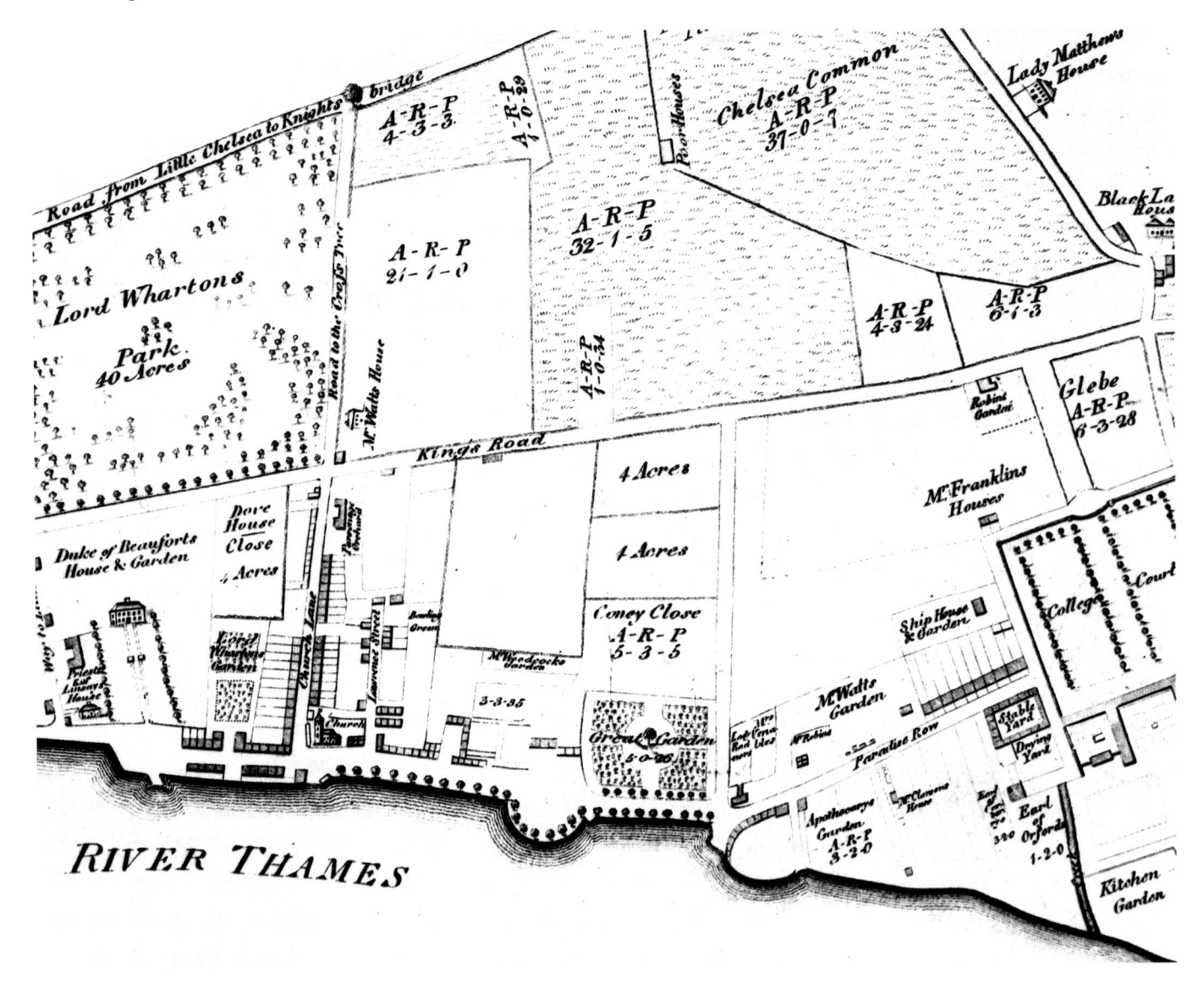

Fig 11 Detail from a map of Chelsea, survey in the year 1664 by James Hamilton continued to 1717 (Royal Borough of Kensington and Chelsea Libraries and Arts Service)

small-scale fine metalworking in the area. Two pits, [61] and [82], produced substantial groups of pottery including complete or near-complete vessels, probably representing the selective clearance of unwanted but still serviceable items.

Pit [61] survived to a depth of 0.40m. It contained 53 sherds of pottery (3810g) from 24 vessels, some complete, dated to 1745–80. In addition to the illustrated pieces (<P1>–<P14>, Fig 13; Fig 14), the tablewares include small sherds from a tea bowl, bowl, two saucers and a plate in Chinese porcelain, some with famille rose decoration. Kitchen wares comprise a post-medieval redware handled bowl (<P9>, Fig 13), a complete Surrey-Hampshire border redware pipkin (<P10>, Fig 13) and part of an English stoneware flagon with long rat-tail handle. Also present are four chamber pots: two in Surrey-Hampshire border redware, one tin-glazed and one in Westerwald stoneware, decorated with applied lion stamps (<P12>, Fig 13). Finally, there are four used crucibles, one complete (Fig 14, < P13>, imported; Cotter 1992), and three broken (Fig 14, <P14>, local).

Pit [82] was excavated to a depth of 0.50m, but may have

been considerably deeper. It produced two fragments of clay pipes with the maker's marks 'GB' <247> and 'IH' <246>. Both were of type AO27 dated to 1780–1820. It also contained six ceramic vessels (552g, Fig 15), including a near-complete Chinese porcelain famille rose teapot <P15>, a tin-glazed mug <P16>, a complete ointment pot <P17> and half a Caughley porcelain toy tea bowl <P18>. Although rare, similar toy tea wares have been found at Finsbury Square (Thomas 2003, 109, no. 8; Blackmore in prep). Also present are sherds of black basalt ware and creamware chamber pot.

Behind Building 3 there were two cesspits, [79] and [81], and two other features that may have been either wells or cesspits, [85] and [874]. All had walls or shafts made of dry laid bricks salvaged from earlier structures. The bricks in features [79], [85] and [874] were of a type manufactured between 1450 and 1700 (fabrics 3033 and 3039). However, the condition of the reused bricks suggests that the features were probably constructed in period 6. Indeed, it is clear that at least three were in use at this time.

Cesspit [79] was lined with walls made of bricks in fabric

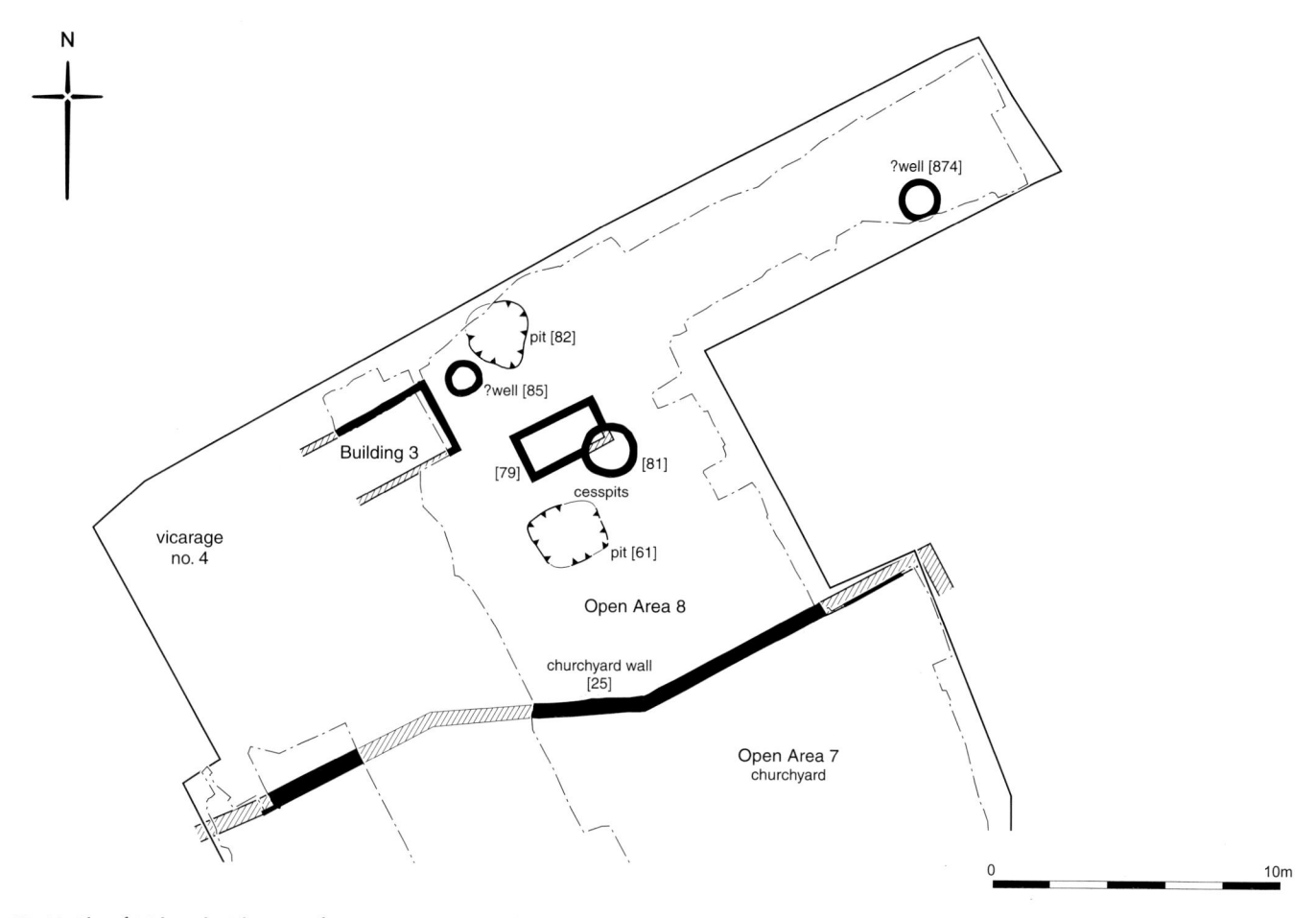

N

?well [874]

pit [82]

?well [85]

Building 3

[79]   [81]

cesspits

vicarage
no. 4

pit [61]

Open Area 8

churchyard wall
[25]

Open Area 7
churchyard

0   10m

Fig 12 Plan of 18th- and 19th-century features in Open Area 8 and Building 3 at 2–4 Old Church Street (scale 1:250)

3039, including many half bats. During the second half of the 18th century it was used for the disposal of a dead cat and domestic pottery comprising a mix of utilitarian redwares and tablewares in creamware, pearlware and English porcelain. Its south-east corner was subsequently truncated during the construction of cesspit [81].

Pottery

The pottery dated to period 6 amounts to 120 sherds (5943g). Over half the assemblage is from the late 18th-century pits [61] and [82], which contained items that were either intact and undamaged, or at least still usable when discarded (Fig 13; Fig 15). Cesspit [79] also contained some sizeable fragments. The scale of the deposits is rather too small to suggest wholesale clearance, although both groups can be quite closely dated and the features were probably not open for long before they were finally filled. Although the reasons for this are unclear, the same was found at OCR97, where a large group of late 17th-/early 18th-century pottery and glass was discarded in a pit (Jarrett 2000b). In the latter group tablewares are in the minority, and other forms dominate (notably tin-glazed wares),

but the reverse applies to the later pits [61] and [82] (dated to c 1745–80 and c 1780–1800 respectively). The two pits and cesspit all contained Chinese porcelain, and pit [62] also contained three crucibles, which may be related to possible mould fragments from OCR97 (Moore 2000, 135, fig 12). Taken together, the pottery from these two sites, and, to a lesser extent CHY96 (Blackmore 1997; Partridge 1997), reflects a period when the area was an elegant and fashionable yet rural place to live (Farid 2000, 118, 126).

Building material

The post-Great Fire assemblage is less elevated in status than that from period 5 and more typical of residential occupation in London; the presence of pantiles suggests that destruction debris from relatively humble buildings such as outhouses or workshops was dumped on the site. The only item of interest is a piece of tin-glazed wall tile (fabric 3067), probably of 18th-century date, which was found in the Wood family burial plot. It has a blue design on a white background showing a figure in a landscape within a roundel, and would have come from an interior of good class.

Fig 13 Late 18th-century pottery from pit [61]: Staffordshire salt-glazed stoneware: tea bowl and saucer <P1>, <P2>, part of a ?sugar bowl <P3>, two tea bowls and saucer with scratch blue decoration <P4>–<P6>; English porcelain bowl, undecorated <P7> (unglazed); Chinese porcelain tea bowl <P8>; post-medieval redware bowl <P9>; Surrey-Hampshire border redware pipkin <P10> and chamber pot <P11>; Westerwald stoneware chamber pot <P12> (line drawings scale 1:4)

Fig 13 (cont)

Fig 14 Crucibles from pit [61]: Hessian stoneware crucible <P13>; Surrey-Hampshire border whiteware crucible <P14> (line drawings scale 1:4)

Fig 15 Pottery from pit [82]: Chinese porcelain teapot <P15> (H 90mm); large tin-glazed mug <P16>; creamware ointment pot <P17>; Caughley porcelain toy tea bowl <P18> with underglaze decoration in blue and factory mark 'S' on the underside (line drawings scale 1:4)

3

Burial practice

3.1 The churchyard in the 18th and 19th centuries

Parish cemeteries

The churchyard of Chelsea Old Church is one of eight recorded burial grounds in the Chelsea area (Table 1). It is the earliest of the three parish cemeteries. By the 1720s the small size of the churchyard had become a problem, leading to the creation of an additional burial ground in King's Road in 1736 (Faulkner 1829, ii, 37). However, as Chelsea continued to grow it became necessary to create further space for burials, so that in 1812 a third parish burial ground was consecrated in Sydney Street, within which the new parish church of St Luke's was built in 1820–4 (Russett and Pocock 2004, 111–13; Fig 1). The parish burial registers list the names of those buried in these cemeteries and the dates of burial. They only occasionally indicate the burial place of specific individuals, but this information may sometimes be gleaned from local histories and, for 1760–70, the parish sexton's book (Kensington and Chelsea Libraries, SR60).

Petyt House

In 1705–7 Richard Ward's school in the north-west corner of the churchyard was rebuilt by William Petyt (Fig 16), a local resident and Keeper of the Records at the Tower of London

Table 1 List of cemeteries in the Chelsea area

Cemetery	Foundation	Area (hectares)	Reference
All Souls Roman Catholic burial ground, Cadogan Terrace, Chelsea	1811	0.61	Holmes 1896; Reeve 1998, 224
Chelsea Hospital Graveyard, Queen's Road (now Royal Hospital Road), Chelsea	-	0.54	Holmes 1896; Reeve 1998, 225
Chelsea Old Church churchyard*	-	0.10	Holmes 1896; Davies 1904, 254–70; Survey of London 1921, 62–79; Reeve 1998, 233
Jewish burial ground, Fulham Road, Chelsea	1813	0.20	Holmes 1896; Reeve 1998, 228
King's Road old burial ground, Chelsea*	1 May 1736	0.30	Faulkner 1829, ii, 7–43; Holmes 1896; Reeve 1998, 229
Moravian burial ground, Milman's Row, Chelsea	-	-	Holmes 1896; Reeve 1998, 229
Royal Hospital Chelsea burial ground (owned by the Royal Hospital)	1692	0.40	Mellor 1985; Reeve 1998, 230
St Luke's churchyard (new church in Robert Terrace, now Sydney Street), Chelsea*	1812	0.91	Holmes 1896; Reeve 1998, 233

*: parish churches

(Faulkner 1829, i, 255–8; Davies 1904, 188–9; Survey of London 1913, 53; *VCH* 2004, 176, 183). The new building, named Petyt House, comprised a covered cloister or arcade of three arches on the street frontage, behind which lay a vestry, with a schoolroom on the first floor and accommodation for the schoolmaster in the attic. Both the cloister and the vestry appear to have been used as a burial ground. Marble tablets on the walls of the vestry commemorated Charity Adams [990], Nicholas Adams [701], Catherine Long [722] and Thomas Long [654] among others (Faulkner 1829, i, 251). Its almost identical replacement, built in 1890–1 (Fig 17), was destroyed by wartime bombing (Chapter 1.1).

Fig 16 Petyt House, built in 1705–7 (from Faulkner 1829, i, facing page 75)

Fig 17 Drawing of Petyt House after it had been rebuilt in 1890–1, by Walter W Burgess (Royal Borough of Kensington and Chelsea Libraries and Arts Service)

The burial population

Before May 1736, when the cemetery in King's Road opened, the churchyard in Old Church Street would have been the burial place for anyone in the parish from wealthy gentry to parish poor. Even after this date it appears that people of varying status were buried there from tradesmen of modest social rank to distinguished professional men. Among their number were bricklayers, carpenters, a barge builder, an apothecary, a butcher, a brewer, a vintner, a pastry cook, a printer, lawyers and army officers.

The excavation revealed 290 burials, including two foetuses. Biographical information on 26 individuals was initially obtained from inscriptions on coffin plates and previously recorded churchyard monuments (Table 2; Fig 18). Coffin plates were found with 93 burials, although they were often extremely fragmentary. Only nine had clearly legible names, ages and dates, and a further 16 had partially legible inscriptions. From this evidence a total of 19 individuals were positively identified from coffin plates, and a further two

(Charity Adams [990] and Robert Butler [462]) were tentatively identified from their position in family burial plots, recorded monumental inscriptions and the relationship of the burial to an individual identified from a coffin plate.

Sequences of burials dated by coffin plates show that 23 individuals died between 1712 and 1842: nine in the 18th century and 14 in the first half of the 19th century. The latest datable burials were those of Richard Gideon Hand and William Wood, who were respectively buried in 1836 and 1842. Another 52 individuals were assigned to date ranges of between about 20 years and 120 years within the 18th century and first half of the 19th century depending on their sequential relationship to the dated burials.

Family histories

Adams family

The remains of Nicholas Adams [701], a bricklayer of the parish, and his first and second wives, Charity [990] and Sarah

Table 2 Excavated individuals for which there is biographical information (coffin types from Reeve and Adams 1993)

Name	Context no.	Age	Date of death	Date of burial	Burial and coffin type
Mr Nicholas Adams	[701]	78 years	7 June 1827	15 June 1827	lead coffin (type 2) in grave in family plot near vestry door
Sarah Adams	[980]	54 or 55 years	26 January 1806	8 February 1806	wooden coffin in grave in family plot near vestry door
Charity Adams	[990]	32 years	1 August 1781	7 August 1781	wooden coffin in grave in family plot near vestry door
Mrs Martha Butler	[430]	84 or 85 years	5 May 1739	10 May 1739	wooden coffin in Butler family burial vault
Mr Richard Butler	[198]	44 years	28 November 1732	2 December 1732	wooden coffin in grave near Butler family burial vault
Mr Robert Butler	[462]	61 years	12 December 1712	19 December 1712	wooden coffin in Butler family burial vault. Burial cost £7 4s
Collon, Collins or [Mary] Cullum (d 1795) (inscription unclear)	[1051]	child	?1797	-	small wooden coffin in grave
Miss Christiana Hamilton	[171]	- months	11 March 1748	22 March 1748	lead coffin (type 2) in grave near the Butler family vault
Mr Gideon Richard Hand	[35]	60 years	13 February 1821	19 February 1821	lead coffin (new type) in grave in family plot
Mr Richard Gideon Hand	[622]	84 years	24 February 1836	2 March 1836	wooden coffin in grave in family plot (on south side of Gideon Hand's grave)
Mr Thomas Langfield	[147]	67 years	5 October 1808	-	wooden coffin in grave
Mrs Catherine Long	[722]	56 years	11 July 1822	18 July 1822	lead coffin (type 1) in brick-lined grave in vestry (Petyt House)
Mrs Esther Long	[1133]	70 years	22 April 1788	-	wooden coffin in brick-lined grave in vestry (Petyt House)
Miss Harriet Elizabeth Long	[719]	19 months	1820	2 February 1820	?wooden coffin, in brick-lined grave in vestry (Petyt House)
Mr John Long	[744]	70 years	1 April 1793	7 April 1793	wooden coffin in brick-lined grave in vestry (Petyt House)
Mr John Long	[713]	68 years	6 November 1822	15 November 1822	lead coffin (new type), brick-lined grave in vestry (Petyt House)
Miss Matilda Long	no skeleton	2 months	30 January 1826	6 February 1826	?wooden coffin [721] in brick-lined grave in vestry (Petyt House)
Mr Thomas Long	[654]	66 years	29 October 1827	6 November 1827	lead coffin (type 2) in brick-lined grave in vestry (Petyt House)
John M[...]	[432]	-	-	13	lead coffin (type 8) in Butler family vault
Mrs Milborough Maxwell	[792]	68 years	10 September 1807	15 September 1807	lead coffin (type 2) in grave
Mr Edward Rainbows	[976]	82 years	1827	23 July 1827	wooden coffin in grave
? Mr Thomas Robson (inscription unclear)	[258]	-	?May 17(3/5)1	-	wooden coffin in grave
Mr Charles Shapley	[525]	70 years	16 September 1780	24 September 1780	wooden coffin in grave
Mrs Ann Wood	[728]	53 years	16 June 1807	22 June 1807	wooden coffin in grave
Mr William Wood	[681]	84 years	22 January 1842	31 January 1842	wooden coffin in grave
? Mary ...od or ...ol	[1055]	?8- years	July 1735	-	wooden coffin in grave

Among the 26 individuals listed here Matilda Long was represented by a coffin plate, thus only 25 individuals were included in the subsample for osteological study

N

vault
[121]

Butler vault

churchyard wall
[25]

C Hamilton

Long family
brick-lined grave
[604]

Wood family

site of Petyt House
1706–1941

Adams family

brick-lined grave
[912]

Open Area 7
churchyard

child (under 17)

Collon

vault
[121]

Butler vault

churchyard wall
[25]

Long family
brick-lined grave
[604]

Wood family
?Mary ?...od

site of Petyt House
1706–1941

Adams family

brick-lined grave
[912]

Open Area 7
churchyard

?Mrs M Maxwell

female/?female

Fig 18 Plans of 18th- and 19th-century burials excavated at 2–4 Old Church Street, showing named individuals or named family vault, burial vaults and the distribution of age and sex (each symbol represents a single individual) (scale 1:150)

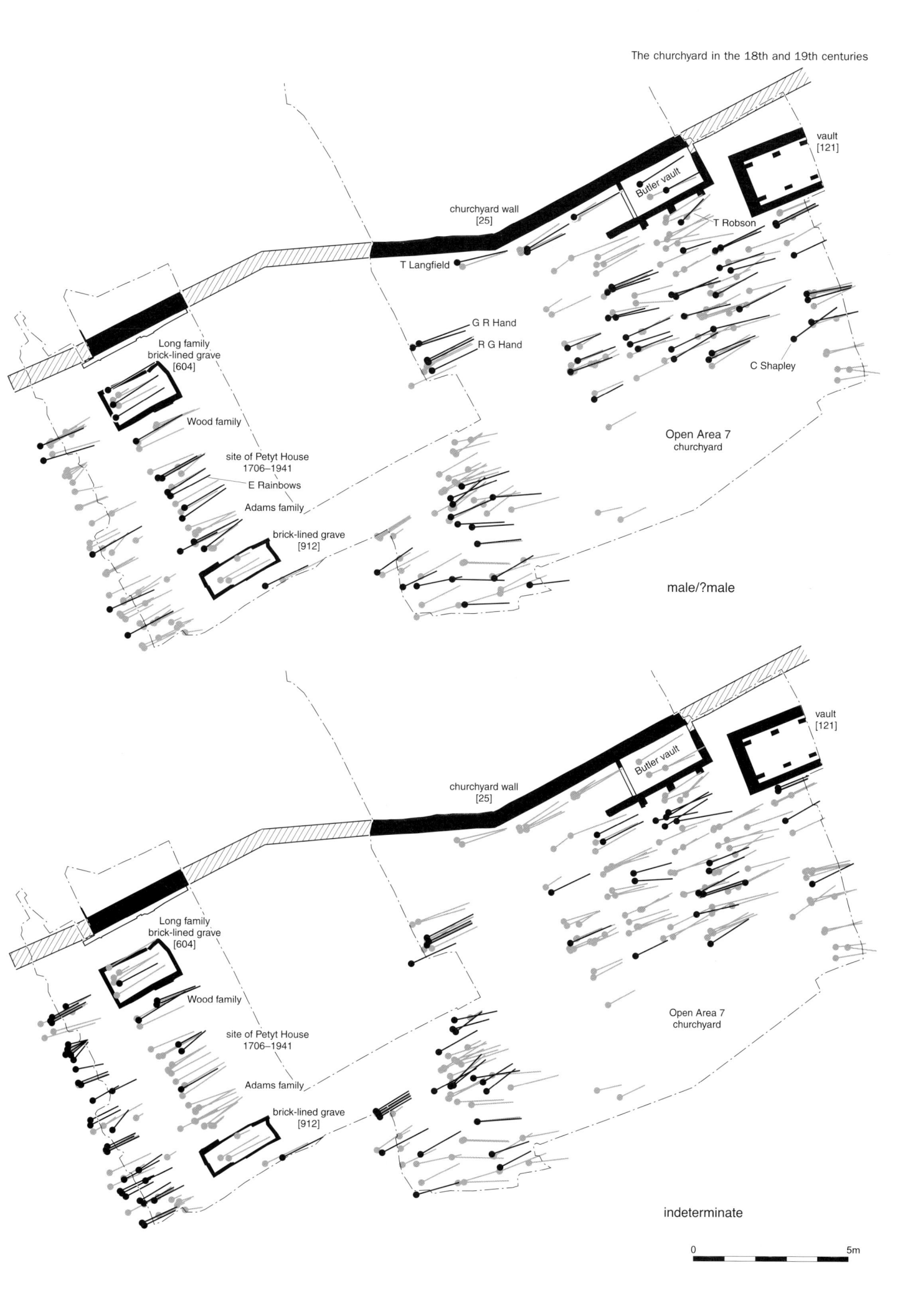

The churchyard in the 18th and 19th centuries

vault [121]

Butler vault

T Robson

churchyard wall [25]

T Langfield

G R Hand

R G Hand

C Shapley

Long family brick-lined grave [604]

Wood family

site of Petyt House 1706–1941

E Rainbows

Adams family

brick-lined grave [912]

Open Area 7 churchyard

male/?male

vault [121]

Butler vault

churchyard wall [25]

Long family brick-lined grave [604]

Wood family

site of Petyt House 1706–1941

Adams family

brick-lined grave [912]

Open Area 7 churchyard

indeterminate

0   5m

[980], were found within overlapping stacks of burials on the west side of the excavation area. In 1816 a flagstone near the vestry door (ie inside Petyt House) marked the burial place of Charity, two infant daughters and Sarah (Chambers 1816, 213). Another flat stone, immediately to the south, marked the burial place of ten of Nicholas's grandchildren, including John Adams (see below). By 1829 a marble tablet had been set up near the vestry door commemorating Nicholas and Charity (Faulkner 1829, i, 251).

Charity's death in August 1781 may have occurred shortly after giving birth, for a neonate [983] in a small lime-filled coffin lay immediately above her skeleton. In May 1782 Nicholas married Sarah Young in St Luke's Church (*IGI*).

At least three of Nicholas's children, John, Charles Thomas and Sarah Elizabeth, survived to adulthood. Sarah Elizabeth, who was christened in December 1774 (*IGI*), married James Faulkner. John, Nicholas's eldest son, was also a bricklayer, who in 1827 resided in his father's freehold property in Wimbledon. John had at least two sons: John (1807–11), who died at the age of 4 years and 6 months (Chambers 1816, 213), and Charles.

In 1798 Nicholas resided at 16 Cheyne Row (Survey of London 1913, 63), but he spent his last years at 14 Milman's Row (later Milman's Street), looked after by his housekeeper Sarah Daley (Fig 1). Nicholas died on the 7 June 1827, an event that he apparently anticipated for he signed his will the previous day (TNA: PRO, PROB 11/1726). He left a substantial estate, comprising freehold properties in Wimbledon and Chelsea. The latter included his home in Milman's Row and property in Lindsey Row (Fig 1). He also left two leasehold estates: one near Chelsea Common and the other in Cheyne Row.

Butler family

Robert Butler ?[462] (1651–1712) was a wealthy gentleman with chambers in the Temple, an estate in Islington Green, which he bought in partnership with Mr Gibbons Bagnall, estates in Wiltshire and Hampshire and property in Chelsea (TNA: PRO, PROB 11/530). The latter included Shrewsbury House (now occupied by 43–44 Cheyne Walk), which he acquired in 1700 and let to various tenants including Robert Woodcock, who used part of the building for his school (Survey of London 1909, 80). Robert Butler also owned two houses in Paradise Row (Fig 1), one of which was Ormonde House, a name that might indicate a familial connection with the Duke of Ormonde who also had the surname Butler (Stephen and Lee 1908, 531–7). The other house was no. 6, where he lived from 1706 until his death (Survey of London 1909, 27).

In 1674 Robert married Martha [430] (Ackinson) in Knightsbridge. On his death 6 Paradise Row passed to Martha, who continued to live there until her death (Survey of London 1909, 27). She bequeathed her substantial estate to her family (TNA: PRO, PROB 11/696). Shrewsbury House, Ormonde House and her home subsequently passed through the family

on her eldest son Edward's side. Ormonde House was taken over in 1779 by Jonas Hanway for use as a school, while Shrewsbury House later became a stained paper manufactory and was eventually demolished in 1813 (Davies 1904, 274; Russett and Pocock 2004, 82–3, 108).

Robert and Martha had at least four daughters: Ann, Elizabeth, Martha and Mary. Martha married […] Good. Mary married Thomas Blow, with whom she had at least three sons, and lived at 6 Paradise Row from 1748 to 1756 (Survey of London 1909, 27). Ann married Gyles Eyre (1673–1739) with whom she had five children: Martha, Elizabeth, Mary, Gyles and Robert.

Robert and Martha also had two sons. Their eldest, Edward (1686–1745), became a Doctor of Civil Law and had a distinguished career serving as President of St Mary Magdalen College, Oxford (1722–45), Vice Chancellor of the university (1728–32) and Member of Parliament (1737–45) (R Darwall-Smith, pers comm). He married Mary Tate (d 1730) with whom he had a daughter, also named Mary (*VCH* 2004, 141). Their second son Richard [198] (b 1688) married Catherine Unsworth at Chelsea on 26 March 1722 (*IGI*; Chambers 1816, 211).

On Edward's death his daughter inherited Shrewsbury House. Mary married twice: in 1747 to Philip Herbert (d 1749), MP, and in 1765 to her first cousin Benjamin Tate. But she had no children, and on her death Shrewsbury House passed to her stepson George Tate (d 1822).

Hand family

At least five members of the Hand family were buried in the churchyard. Three are known from monumental inscriptions: Mr Richard Hand, who died in April 1767 aged 65, his wife Margaret, who died in July 1790 aged 73, and their son George Hand, who died in 1769 aged 6 years and 3 months (Chambers 1816, 209). The remains of George's elder brothers, Gideon Richard Hand [35] and Richard Gideon Hand [622], were found during the excavation and identified from coffin plates. Margaret's niece, Elizabeth Mercer (d 1818), was also buried in the churchyard (ibid).

Four generations of the family, credited as creators of the Chelsea bun, owned and ran the Chelsea (or Royal) Bun House. The Bun House (Fig 19) was situated on the north-west side of Grosvenor Row (now Pimlico Road) and was well placed to take advantage of passing trade, especially holidaymakers who resorted to the adjacent Five Fields (now Belgravia) in their thousands (*The Builder* 1901; Blunt 1921, 54–6). Business was particularly brisk on Good Fridays, when large crowds would mob the house. On one such occasion, in 1793, so many were gathered outside that Mrs Hand had to announce that they would not be selling hot cross buns that day, only the ordinary Chelsea buns. Although trade declined after the closure of Ranelagh Gardens in 1804 the Bun House continued to attract considerable custom during holidays, and it is recorded that on Good Friday in 1829 240,000 buns were sold (Blunt 1921, 55; Denny 1996, 38).

Fig 19   View of the Chelsea Bun House shortly before its demolition in c 1839 (Royal Borough of Kensington and Chelsea Libraries and Arts Service)

The Bun House had achieved renown by the end of the 17th century, and at the height of its popularity in the 18th century it was frequented by royalty including George II and George III. On one occasion George III presented Richard Hand with a silver half-gallon mug containing five guineas (Blunt 1921, 55). The Hand family made the most of this patronage (Fig 20). An insurance policy for the 'Royal Bunhouse', was taken out in February 1781 by Margaret Hand who is recorded as 'a pastry cook' (Sun Fire Office Policy no. 439073).

The last proprietor, Richard Gideon Hand [622], cut a somewhat eccentric figure habitually attired in a long dressing gown and Turkish fez. In his younger days he served as a Lieutenant in the Staffordshire Militia from 1793 to 1808 (Militia List: National Army Museum, pers comm 2005) and was somewhat endearingly known as 'Captain Bun' (Blunt 1921, 56). He died in 1836, and the Bun House was demolished 3 years later.

Thomas Langfield

Thomas Langfield's [147] last home was in Danvers Street (Fig 1), where he lived with Mary Newman, a spinster, who was an executrix of his estate. He had family connections with Somerset, where his brother and half sister, John and Betty Langfield, lived in the parish of Stokeunderham (Stoke sub Hamdon). His nephew, Joseph Langfield of Hatfield Broad Oak

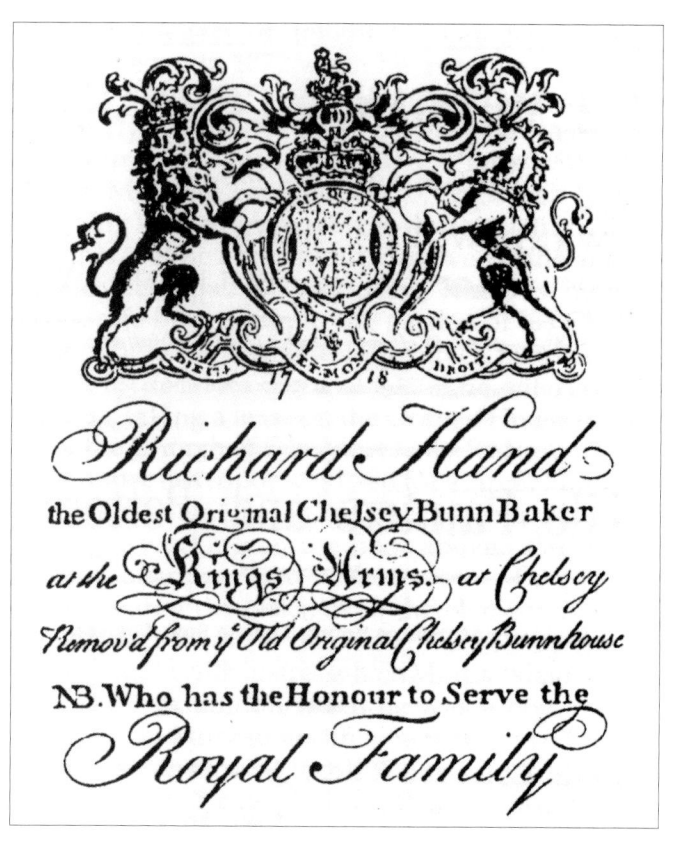

Fig 20   The trade card of Richard Hand of the Chelsea Bun House (Royal Borough of Kensington and Chelsea Libraries and Arts Service)

in Hertfordshire, and Richard Burnard (surgeon) of Crookham in Somerset were the other executors of his estate of under £5000. All were beneficiaries in his will, as were Ann and Mary Grant, and Hannah Latteny, wife of Josias Latteny of Clerkenwell, and her daughters Mary and Harriet (TNA: PRO, IR 26/140; PROB 11/1486).

Long family

A brick-lined grave under a flat stone in or near the vestry room (in Petyt House) contained seven individuals from four generations of the Long family (Chambers 1816, 205). John [744] (d 1793) and his wife Esther [1133] (d 1788), who lived in the nearby hamlet of Little Chelsea on the Fulham Road (Fig 1), represented the first generation. They had a daughter, Esther, and two boys. Their eldest son, John [713] (d 1822), was a wealthy gentleman, who owned three properties in Beaufort Row (Fig 1): nos 9 and 10, and his home at no. 18.

Their other son, Thomas [654] (d 1827) and his wife Catherine [722] (d 1822), lived at 56 Upper Sloane Street. Both were commemorated on a marble tablet in the church (Survey of London 1921, 19). Shortly after Catherine's death Thomas moved to 18 Beaufort Street, which he inherited from his brother. Thomas and Catherine had five sons and five daughters.

One son, John (b 1792), lived in Beaufort Place (Fig 1) and was a carpenter by trade. He married Sarah Waite, and in June 1818 they had a daughter Harriet [719], who died 19 months later. Sarah and John had at least three more children: Henrietta (b 1825), Oliver (b 1826) and Sarah (b 1832) (IGI).

One of John's brothers, Thomas, was a coal merchant who lived in Beaufort Place and inherited a property in Duke Street from his father (Fig 1). Thomas's daughter, Matilda [721], was born in December 1825, but died 2 months later.

Charles Shapley

An inscription on a broken flat stone, which was removed from the churchyard by the sexton in 1822, provided some biographical details about Charles Shapley [525] (Chambers 1816, 212). It mentioned Charles, his first wife Ann (d 1742), his third wife Grace (d 1769), his daughter Sarah and Mary Shapley wife of ?Jn Shapley (possibly his son). He had at least four children with Grace: Sarah, Susannah, Catherine and Fanny, all of whom were christened between March 1750 and June 1757 (IGI).

Wood family

A discrete burial plot belonging to the Wood family, immediately to the south of the brick-lined grave of the Long family (Fig 18), was found to contain the remains of four adults. Two were identified from coffin plates as William Wood [681] (d 1842) and his wife Ann [728] (d 1807). The others were unidentified; one [1134] had been buried before Ann, and the other [697], a woman over 46 years old, was buried after Ann but before William.

Although William Wood was not born in Middlesex he spent most of his long life in Chelsea, and in 1792 he was listed in a trade directory among its principle inhabitants. He was a butcher by trade, but as a trusted member of the community he was appointed by the vestry as a beadle of the parish. His duties as beadle, a minor parochial servant, would have included assisting other parish officials in their work and keeping order in church, for which he received a salary of about £70 per annum. In 1827/8 the annual accounts for the parish list the salary for two beadles as £140 (Faulkner 1829, ii, 33).

He and Ann had at least four children: Mary (b c 1784), Ann Elizabeth (1786–9), who died at the age of 3 years and 2 months, William Henry (b c 1791) and Elizabeth (b 1793), who died at the age of 4 months. Mrs Wood and her two infant daughters were commemorated on the monument above their grave (Chambers 1816, 205), although the remains of the infants were not found during the excavation. Her surviving daughter, Mary, married Thomas Boyland, and their son, William Henry, became a painter and glazier.

William lived for 35 years after the death of his wife, but apparently never remarried. In later years he was looked after by his housekeeper, Elizabeth Shrives, who is mentioned in his will of May 1835 (TNA: PRO, PROB 11/1958). At the time of the census in 1841 he was living with his daughter Mary, who was by then a widow, his son William, and a 50-year-old servant, Johannah ?Patery, at his home in 3 Church Lane. From his room, on the first floor at the front of the house, he would have had a good view of the nearby church. He apparently died of old age, for his death certificate simply records 'decay of nature' as the cause of death (TNA: PRO, IR 27/263). He left his home in Church Lane and a freehold property near the Orange Tree in Richmond, Surrey, to his son.

Layout and appearance of the churchyard

There are numerous 18th- and 19th-century pictures of the church and the surrounding area, but nearly all are from the south or east and none include the churchyard on the north side of the church (Longford 1980). Nevertheless, detailed descriptions of the monuments in the churchyard appear in a number of publications (Lysons 1795; Faulkner 1829, i, 242–53; Davies 1904, 254–70; Survey of London 1921, 62–79) and in Chambers's (1816) manuscript in the Royal Borough of Kensington and Chelsea Central Library. These sources include transcripts of inscriptions containing biographical information about some individuals exhumed during the excavation. Chambers's manuscript was originally accompanied by a plan of the monuments, but this is now missing. However, Walter Godfrey's schematic sketch plan of the burial ground for the *Survey of London*, although not to scale, shows the relative position of burial plots and indicates those with inscribed monuments (Fig 21; Survey of London 1921, 62). Although not accurate enough to allow close correlation with the excavation plans, it indicates the approximate position within the excavation area of burial plots belonging to specific

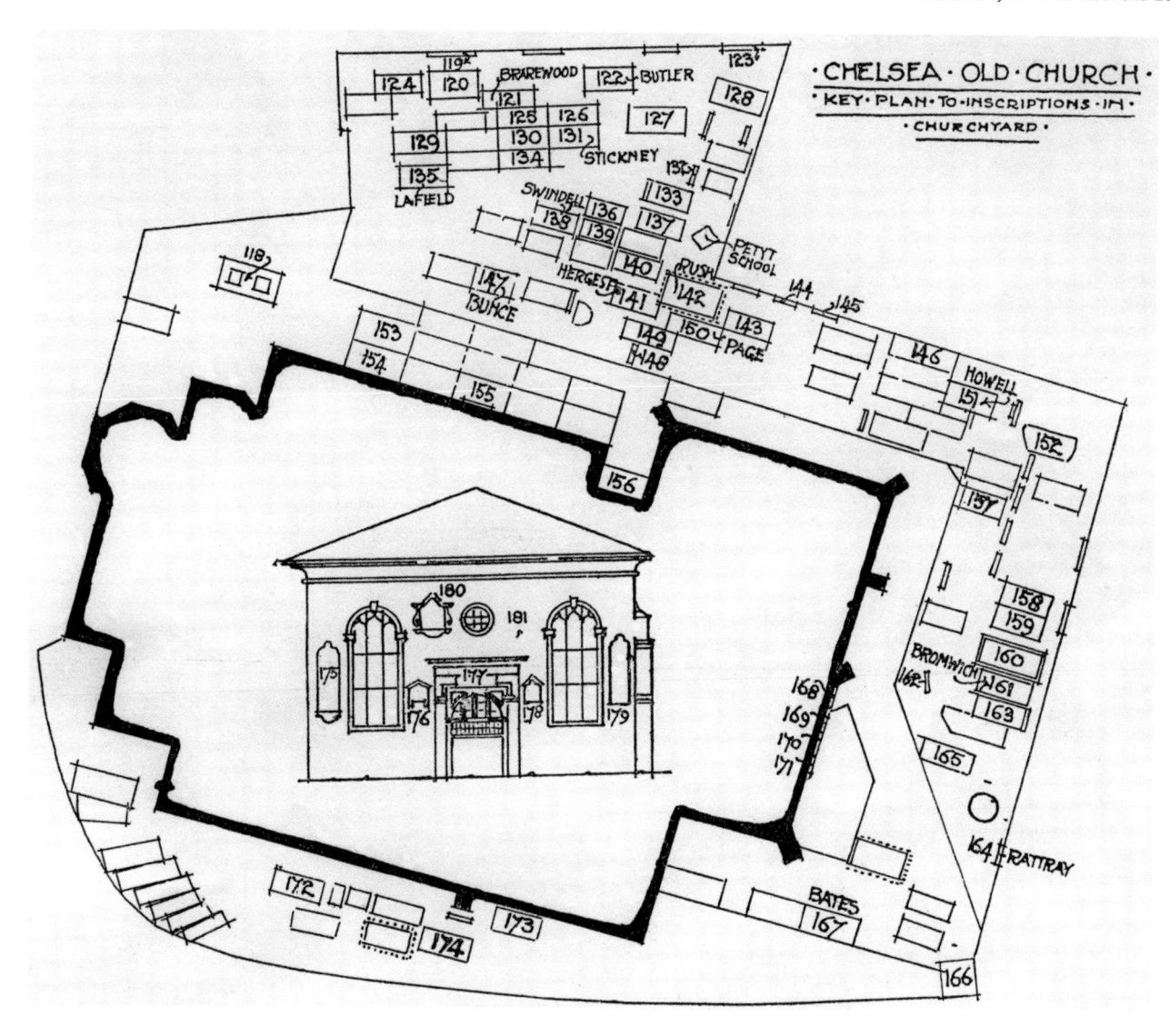

Fig 21 Schematic plan of monuments in the churchyard by Walter Godfrey (from Survey of London 1921, 62)

individuals and family groups.

Most of the monuments extant at the time of the 1921 survey comprised flat slabs (Survey of London 1921, 62–79). Of those dating up to 1800, 40 were flat slabs or ledgers, including one on a raised kerb, and six were headstones. A number of more substantial monuments were located around the edge of the churchyard. These included five table monuments, two of which lay within the excavation area – one above the Butler family vault and the other over a vault containing Mrs Sarah Eyre. Of the few monuments that survive today the most imposing is that of Sir Hans Sloane (1660–1753; Fig 22), founder of the British Museum and president of the Royal Society and of the College of Physicians (Faulkner 1829, i, 338–73; Blunt 1921, 97–139). It is situated to the south-east of the church, and comprises a pedestal tomb in Portland stone surmounted by a large funeral urn and canopy.

Many of the burial plots in the churchyard were aligned, according to convention, approximately west–east, parallel to the church (Fig 23). However, the most northerly ones, including most of those in the excavation area, generally lay parallel to the north wall of the churchyard on a west-south-

west–east-north-east alignment. This arrangement was probably designed for neatness and to make economic use of available space.

Although there had been burials at Chelsea since medieval times and the parish register records burials dating back to 1559, the northern part of the churchyard does not appear to have been used as a cemetery before about 1700. The excavated burials generally appear to date to either the 18th century or the first half of the 19th century, although a few might possibly date to the last years of the 17th century. Many graves cut through late 17th-century quarry pits (OA6, period 5; Fig 10). Thus it seems that the burials spanned a period of about 150 years.

Records of monumental inscriptions indicate that earliest burials in the excavation area lay next to or near the churchyard wall, and included those of John Pennant (d 1709), Robert Woodcock (d 1710), Robert Butler (d 1712) and probably Honour Francis (d 1695). Coffin plates show that some of the latest burials were those near the street frontage, either in or next to Petyt House, including members of families with the surname Adams, Long and Wood (Fig 18).

Fig 22 *View of Cheyne Walk from the churchyard of Chelsea Old Church, with the tomb of Sir Hans Sloane on the right, by William Parrott 1840 (Guildhall Library, City of London)*

Graves

The graves were generally arranged in rows, of which there were at least ten. Due to overcrowding the ends of graves in one row would often overlap those in the next (Fig 23; Fig 24). The two most clearly defined rows were in the north-west corner, where the presence of Petyt House would have necessitated greater regularity. Burials were often stacked several deep within a plot.

Brick-lined graves

There were two brick-lined graves at either end of a row of burials on the west side of the site (Fig 18). Both had been filled with earth.

One [604] belonged to the Long family, which according to parish records lay under a flat stone at the north-west corner of the vestry room, that is within Petyt House. It contained the stacked burials of seven members of the Long family. The earliest was that of Esther [1133], followed in succession by John [744], Harriet [719], Catherine [722], John [713], Matilda [721] and Thomas [654]. Catherine, John and Thomas had been placed in lead-lined coffins. Artefacts recovered from the grave included a complete wig curler and part of a whittle knife. Both dated to the 18th century and were probably residual. However, a gold-coloured ring (possibly part of an earring) may have belonged to one of the interred (below, 3.5).

The other grave [912] was rectangular in plan, and was 2.5m long and up to 0.94m wide. Its poorly constructed walls and roof were made of a mixture of red brick (fabric 3046), dated to c 1450–1700, and more recent yellow stock brick. The

north wall was 0.71m high, but the south wall was 0.14m lower, giving the structure a curious lopsided appearance. A single lead-lined wooden coffin [919] lay inside the grave (Fig 25); it contained the remains of a woman over 46 years old.

Burial vaults

Only two structures could be strictly defined as burial vaults, that is a subterranean chamber capable of accommodating two coffins side by side (Litten 1991, 207).

The earlier of the two was almost certainly the Butler family vault, which adjoined the churchyard wall (Fig 18; Fig 26). It was a well-made structure comprising a rectangular chamber with floor, walls and barrel-vaulted roof made entirely of red brick. The interior of the chamber was 2.33m long, 1.44m wide and 1.36m high (Fig 27). The vault floor extended through an entrance on the west side into a small external area flanked by walls to the north and south. Both the north wall and the vault were keyed into the churchyard wall, suggesting that they may all have been contemporaneous. If so it is possible that the Butler family paid for the churchyard wall. Prior to the burial of Martha Butler [430] in 1739 part of the roof had either collapsed or been broken, and was crudely patched with brickwork (Fig 26).

The chamber contained the remains of three individuals. The earliest interment is assumed to be that of Robert Butler [462], as indicated by the inscription on the table monument that once overlay the vault (Survey of London 1921, 64). It had been placed in a wooden coffin on the floor on the south side of the chamber. The remnants of the eastern half of the coffin

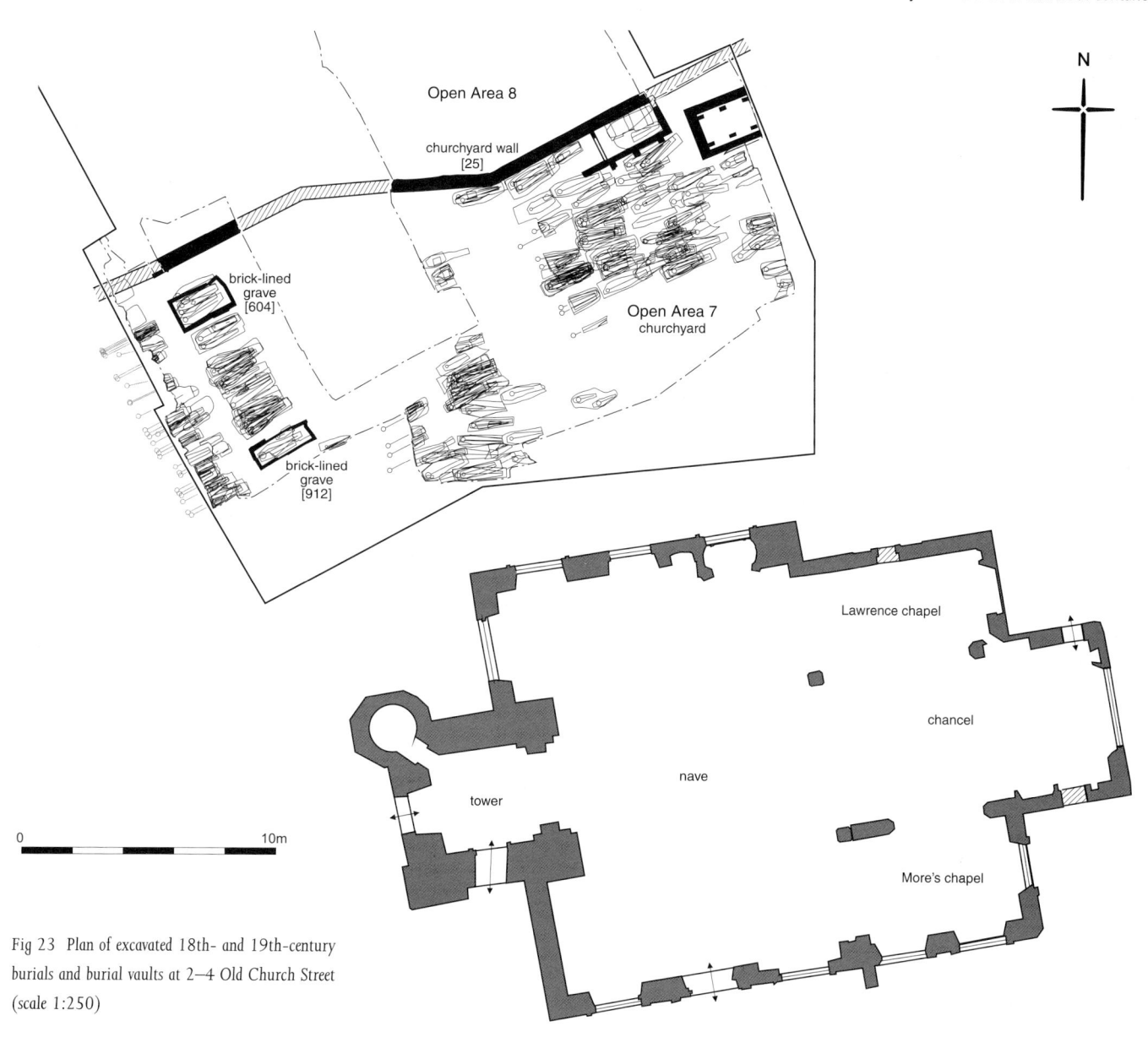

N

Open Area 8

churchyard wall
[25]

brick-lined
grave
[604]

Open Area 7
churchyard

brick-lined
grave
[912]

Lawrence chapel

chancel

nave

tower

More's chapel

0   10m

Fig 23 Plan of excavated 18th- and 19th-century
burials and burial vaults at 2—4 Old Church Street
(scale 1:250)

Fig 24 Burials under excavation,
looking north

29

Fig 25 The lead-lined coffin inside brick-lined grave [912], looking west

Fig 26 The Butler family vault with blocked entrance at west end and crudely patched roof, looking east

were very badly decayed and lay with skeletal remains beneath a pile of rubble from the collapsed roof. The western end of the coffin and much of the upper part of the skeleton had been truncated, probably during the removal of some of the rubble. The remaining debris had been flattened to provide a level base on which to lay the wooden coffin of Martha Butler [430], which was identified from the inscription on a lead coffin plate. The third burial lay in a lead coffin, which rested on three roughly built support walls abutting the north wall of the vault. The latter were 0.44–0.48m high and extended south to the middle of the chamber. The coffin contained the skeleton of a man over 46 years old [432]. The name on the coffin plate was partly decipherable as 'John M[…]'.

Vault [121] lay near the north-east corner of the churchyard (Fig 18) and, judging from Godfrey's plan (Fig 21, Survey of London 1921, 62–3), was possibly that of Mrs Francis Eyre (d 1756), which was marked at ground level by a table monument. The walls of the vault were made of red brick (fabric 3032) and enclosed an empty chamber that was 2.10m long and 1.48m wide. Its floor and slightly arched roof were also of brick. The roof had apparently been rebuilt during repairs, when the east end of the vault was underpinned with concrete. The vault was emptied probably during this work.

Fig 27 Interior of the Butler family vault, looking west

3.2 Coffins

Adrian Miles

Coffin types

Most coffins were wooden and only survived as a dark brown organic deposit, although nine lead coffins were also found. The latter conformed to the standard 18th- and 19th-century 'triple shell' pattern of a wooden inner coffin sealed in a lead shell, which was then covered with a second wooden outer coffin (eg Fig 28). This was the main decorated coffin visible at the funeral, with the decorative name plate, lid motifs, escutcheons, grips and grip plates.

The lead coffins were recorded using the construction types identified during the Christ Church Spitalfields excavations (Reeve and Adams 1993, 84–5). The divisions of types can be seen in Table 3, although two did not conform to those previously recorded.

At Chelsea Old Church it was not possible to identify the type of wood used to make the coffins, but on other sites from the same period, where it was possible, it was shown to be elm (Miles 1994; Brickley and Miles 1999). Elm was used as it was less likely to split being cross-grained, and was to some extent water resistant (Litten 1991, 90). Coffins were normally constructed from six pieces of wood; two sides, two ends, a base and lid. From the better preserved ones, such as that of Martha Butler [431], it is possible to see that the sides have up to six kerfed saw lines internally to allow the sides to be bent to fit the base. Generally they were nailed together, although some were also screwed together. In some instances from Chelsea Old Church the internal base of the coffin was sealed with resin or pitch, which was then covered with sawdust or bran, which was up to 20mm thick in [431]. Sometimes rosemary or balm was added to counteract the smell of decomposition. This not only provided a soft base to lay the body on but also acted as an absorbent for the body fluids (Plume c 1910, 21).

Most of the coffins were decorated with upholstery studs of *c* 10mm diameter. These were found in both copper alloy and black painted or enamelled iron. Different patterns were noted, the majority of which were based on a double row of studs around all the edges of the coffin, although single rows were common and occasionally triple rows. Due to the state of

<table>
<tr><td colspan="3">Table 3 Lead coffin construction types (types from Reeve and Adams 1993, 84–5)</td></tr>
<tr><td>Context no.</td><td>Coffin type</td><td>Date</td></tr>
<tr><td>[36]</td><td>new type</td><td>1821</td></tr>
<tr><td>[172]</td><td>2</td><td>1748</td></tr>
<tr><td>[433]</td><td>8</td><td>?1813</td></tr>
<tr><td>[653]</td><td>2</td><td>1827</td></tr>
<tr><td>[702]</td><td>2</td><td>1827</td></tr>
<tr><td>[714]</td><td>new type</td><td>1822</td></tr>
<tr><td>[723]</td><td>1</td><td>1822</td></tr>
<tr><td>[793]</td><td>2</td><td>1807</td></tr>
<tr><td>[919]</td><td>2</td><td>?</td></tr>
</table>

Fig 28 Coffin of Thomas Long [654] in brick-lined grave, looking west

preservation of the coffins it was not possible to determine with any certainty how many patterns were used. Although primarily decorative, they were also used for holding the fabric covering in place on the outside of the coffin (below, 3.3).

It is reasonable to assume that the level of decoration on a coffin is a good indicator of the status of the individual, or at least of the person paying for the funeral. A contemporary trade booklet (Turner 1838) shows that the cost of the coffin varied with the number, type and quality of the items attached.

Coffin plates

Only lead plates survived in any recognisable form, 20 of which were recovered. These form two distinct groups: those attached to the inner lead coffin and those from the wooden outer. There were nine from the lead shells and 11 outer plates. Overall, they represent a typical assemblage for the period.

The plates attached directly to a lead shell are normally plain, often with an inscribed or punched border, and were used to identify the coffin. The outer plate was the main decorative plate, which was intended to be seen at the funeral. These were produced with a variety of designs, examples of which have been recorded from other archaeological excavations. To date, the only published collection from London is from Christ Church Spitalfields (Reeve and Adams 1993), which is the main source for comparative material. Three trade catalogues for the period also survive in the Victoria and Albert Museum, London. These are the 1783 catalogue of 'J.B.' issued through Tuesby & Cooper of Southwark (V&A acc no. E997 to E1011–1903 (M 63e)), the c 1821–4 catalogue by 'A.T.' (V&A acc no. E994 to E1021–1978) and the 1826 catalogue by 'E.L.' (V&A acc no. E3096 to E3132–1910).

The preservation of the plates is generally poor, with only two in good condition. The remainder all had some damage or corrosion, which makes comparison with other material difficult.

Of the 11 decorative plates (Table 4), two are too badly damaged to allow comparisons to be made. The remaining nine all have parallels in the Christ Church Spitalfields material (Reeve and Adams 1993). There are also parallels to be found within the unpublished collections recorded at New Bunhill Fields, Islington Green (IGN96; Miles 1997) and St Andrew Holborn (HUD01; Miles 2002). None of the plate designs from Chelsea Old Church appear in the trade catalogues.

Lead plates are much more varied than the cheaper stamped tin-dipped iron plates. The raised borders in particular tend to be individually hand chased. Although identical plates are found, it is much more common to find plates that have the same general appearance, but on closer examination have variations in the foliage, scallop shells, flowers or many other features. Therefore, the comparisons are based on a general style, such as one <5>, which is similar to type 46, but has a different style of flower at the top, with the stem upwards. However, a much closer parallel was found at St Andrew Holborn, dated to 1828, but adult sized. Another, <242>, is similar to type 14, but the leaves and flowers are different and the border for the inscription field is also different. Again, St Andrew Holborn provides a much closer comparison, dated to 1791. One plate <13> has no direct comparison with the Christ Church Spitalfields material, but the level of damage and corrosion to the plate makes it difficult to be certain what the nature of the raised decoration at the top centre is. If it is a shield, then the closest parallels are type 8 or type 86.

The raised shield and/or flower motifs most commonly found at Chelsea Old Church (<4> (Fig 29), <5>, <12>, <13>, <122>, <164>, <175> and <242>) are typical of an early 19th-century lead plate assemblage. The variations consist of normal shields or florid roccoco shapes, while there is a large degree of stylisation of the flowers and leaves. The flowers are normally roses, either as full blooms or as buds.

Only one plate does not conform to this general pattern (<3>, Fig 29); this features a seated angel with trumpet, urn, roccoco shields and cherubs. It is a reasonably common design, with examples from Christ Church Spitalfields in both lead and iron, St Andrew Holborn in lead (1774–83) and in iron from Islington Green (1825–49).

The nine plain lead plates from the site form two groups: those with only an inscription (<2>, <10>, <128>, <172> and <173>) and those that also have punch-stamped or incised decoration. Apart from one <9>, which has stamped diamonds

Table 4 *Types of outer decorative coffin plates (types and date ranges from Reeve and Adams 1993)*

Coffin no.	Acc no.	Skeleton no.	Name	Sex	Age	Year of death	Type	Fig no.	Date range
[+]	-	-	-	female	adult	?179-	8	-	1767–1825
[146]	<122>	[147]	Thomas Langfield	male	adult	1808	8	-	1767–1825
[431]	<174>	[430]	Martha Butler	female	adult	1739	damaged	-	-
[526]	<3>	[525]	Charles Shapley	male	adult	1780	9	Fig 29	1773–93
[623]	<4>	[622]	Richard Gideon Hand	male	adult	1836	21	Fig 29	1824–47
[653]	<12>	[654]	Thomas Long	male	adult	1827	82	-	1820–9
[714]	<13>	[713]	John Long	male	adult	1822	no clear comparison	-	-
[721]	<5>	-	Matilda Long	female	child	1826	?46	-	1786–1821
[723]	<175>	[722]	Catherine Long	female	adult	1822	82	-	1820–9
[729]	<164>	[728]	Ann Wood	female	adult	1807	similar to type 82	-	1820–9
[1132]	<242>	[1133]	Esther Long	female	adult	1788	?14	-	1743–1818

<3>

Fig 29 Coffin plates: from the coffin of C[harles] Shapley <3> (H 420mm); from the coffin of Richard Gideon Hand <4> (W 290mm at top) (cont overleaf)

<4>

Fig 29 (cont)

flanking the year of death, decoration is used to form a border around the inscription field. One plate <176> has two punched lines of hearts around the edge, with the hearts touching at their top and bottom. Two plates, <11> and <163>, have the same punched design, a primitive flower/palmette, but the direction of the stamp has been reversed between the two plates.

Coffin handles

Coffin handles recovered from other MoLAS excavations of the period, such as St Benet Sherehog (ONE94; Miles and White in prep) and St Bride's lower churchyard (FAO90; Miles 1994) have been split into six different types, four of which are found here (Table 5). Type 1 is a simple, plain rounded style with no embellishments. Type 2 is also a simple rounded style but with

a swelling in the centre of the grip. The variations in the size of the swelling have not been taken into account. Type 3 is a plain right-angled handle; type 5 uses winged cherubs or other figures to form the grip.

Table 5 Types of coffin handle

Type	No.	% of recorded handles	Date range from Chelsea Old Church
1	24	26.4	-
2	18	19.8	1780–1842
3	36	39.6	1735–9
5	13	14.2	1807–36
Total	91	100.0	

Although precise dating is impossible for the vast majority of the coffin handles from the site, a number of things can be stated with reasonable certainty. The type 3, right-angled handle is the earliest dated form, which is found on coffins dated from 1735 through to 1739 on this site. This type has also been found in contexts dated 1687–1720 at St Paul's Cathedral (SAT00; Wroe-Brown 2001). Julian Litten (pers comm) also suggests that these only date between *c* 1650 and *c* 1750.

The rounded types 1 and 2 proliferate through the latter part of the 18th and the first half of the 19th century, while of the decorated type the author has only found five in 19th-century burials.

3.3 Coffin textile furnishings

Nicola Powell

Coffin coverings

Prior to the introduction of wood finishes such as French polishing in the early to mid 19th century, it was customary to enhance the appearance of the outside of a coffin with a cloth covering such as wool or silk (Janaway 1993, 95). The evidence for textiles from the graves at Chelsea Old Church is fragmentary in general and little for coffin coverings. However, where they do survive, they may indicate what may have been common practice or a way of showing a high social standing in life as well as after death (Fig 30).

The coffin [431] of Martha Butler [430], produced part of the velvet coffin covering. Fragments from another coffin [284] <231> and <232> covering are likely to be wool. All are discoloured brown, but may once have been a vivid scarlet, midnight blue, holly green, turquoise or black (Janaway 1993, 95).

All the fragments discussed above showed evidence of the copper-alloy pins or studs that would have fixed the covering in place (above, 3.2). They were usually arranged in a pattern, serving a decorative as well as functional purpose (eg <226>, Fig 30).

Coffin linings

As the coffin was left open for a few days prior to interment so the body could be viewed and last respects made, the interior of the coffin had to be tastefully decorated with linings and frills. Mrs Milborough Maxwell's coffin [793] contained decorative textiles that may have come from the inner coverings of the coffin or other funerary textiles such as a pillow covering or winding sheet (<229>, Fig 30). They are made of wool, discoloured to a light brown and decorated with a pinked cut edge and punched decoration. Coffin lining was also recovered from a lead-lined coffin [919] in brick-lined grave [912] and a piece of folded wool, probably lining (<226>, Fig 30), from coffin [452] in grave [45].

Martha Butler's coffin [431] contained a quilted silk lining secured with tacks, two pieces of which were recovered (<225>, Fig 30). Both are quilted with a plain diamond hand-stitched pattern. One fragment is from the inner covering of the foot of the coffin lid and has been simply turned down to form a hem and roughly stitched. The implication is that as long as the overall effect was of top-quality materials and workmanship, the inside of the coffin did not warrant closer inspection. A coarser fabric, such as wool, was then used to line the quilted silk and a small part of this remains along one of the hems.

A number of copper-alloy ball-headed pins were recovered, some large, which may have attached the coffin linings to the wooden coffin and finer ones, which may have secured funerary clothing (below, 3.4).

3.4 Funerary clothing and shrouds

Nicola Powell

The fragmentary nature of the textiles recovered from the graves makes it difficult to be certain what may be covering the body or the inside of the coffin for burial. Before the Act of Parliament of 1660 that forbade the use of anything but wool for shifts, shrouds and winding sheets was repealed in 1815, there was no choice of material for burial clothes (Janaway 1993, 95). The use of wool supported the woollen trade, however, on payment of a £5 'fine', more delicate fabrics such as silk could still be used by those who chose to pay, presumably the wealthy. Winding sheets were still used up to the 19th century and were decorated with punched designs. It may well be that the wool fragments from Mrs Milborough Maxwell's coffin [793], with the pinked edging and punched decoration could be a winding sheet (<229>, Fig 30). It is interesting to note that Mrs Maxwell was buried before the repeal of the Act of Parliament 1660. Small fragments of material (made of wool) from other coffins [323], [396], [452] and [954] could equally be evidence for funerary clothing, garments or coffin linings but may be indistinguishable.

A number of pins of varying size were recovered (Fig 31), all came from coffins or the fills of graves. The finer ones, such as a complete pin from one skeleton [1048], may have been used to fasten shrouds or winding sheets. Otherwise, funerary clothes were fastened with tapes.

Five small copper-alloy rings were recovered from graves, four from one grave [670] (eg <157> and <160>, Fig 32) and one from the grave of William Wood [681], who died in 1842. It is not clear what purpose they served, but it seems likely they are associated with funerary clothing or furnishing. A sixth ring was found with another skeleton [775] (<249>, Fig 32). Also recovered was a small hook (<248>, Fig 32) from the same grave, part of a hook-and-eye fitting, still familiar today, as a cheap and effective clothing fastening.

<225>

<229>

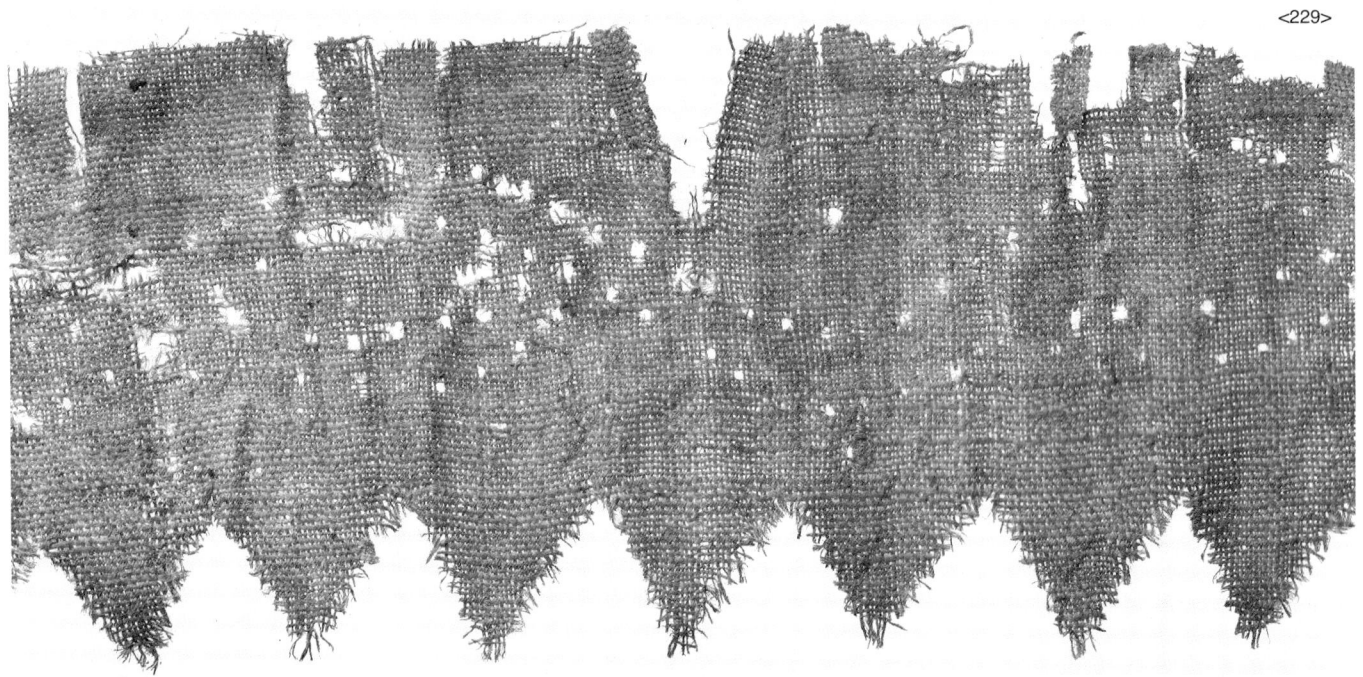

Fig 30 Coffin furnishings: silk <225> from the quilted lining of Mrs Martha Butler's coffin (W 75mm); pleated edging <229> from Mrs Milborough Maxwell's coffin, possibly from a pillow or inner covering (W 156mm); velvet or felt <226> from outer covering of coffin (W 176mm)

<226>

Fig 30 (cont)

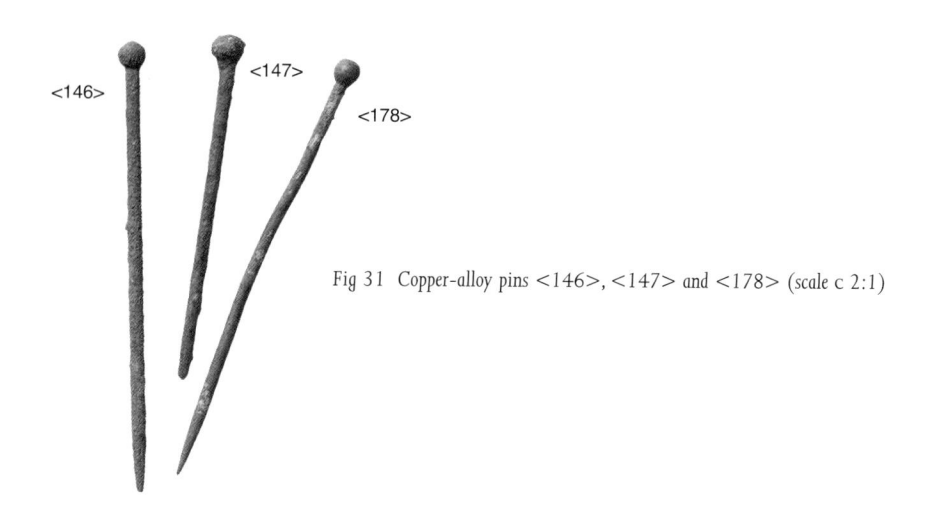

<146>

<147>

<178>

Fig 31 Copper-alloy pins <146>, <147> and <178> (scale c 2:1)

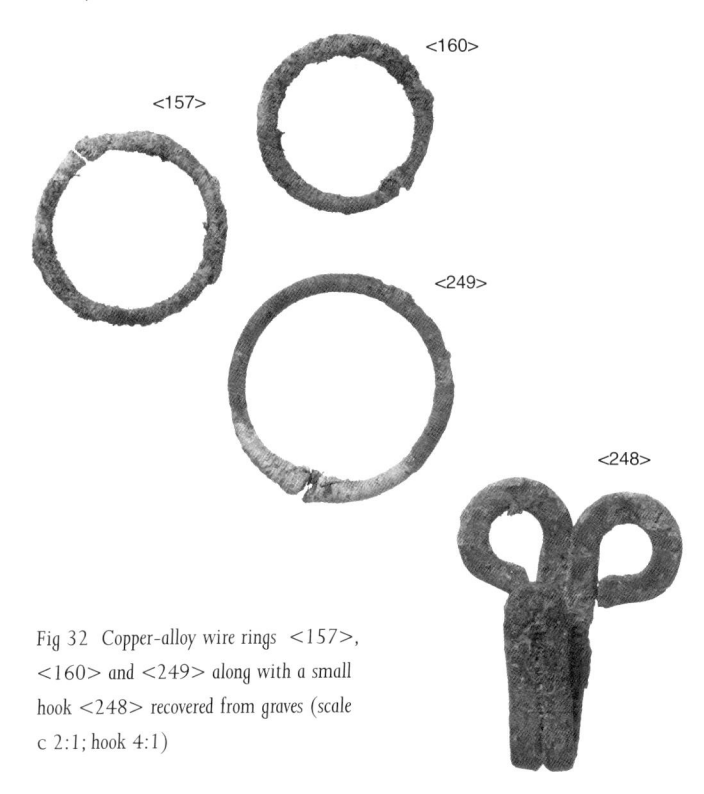

Fig 33 *Small wire loop with twisted ends <148>*
(scale c 4:1)

Fig 32 *Copper-alloy wire rings <157>,*
<160> and <249> along with a small
hook <248> recovered from graves (scale
c 2:1; hook 4:1)

A small wire loop with twisted ends was found in a pit [327] (<148>, Fig 33). Such objects are commonly found in late medieval and early post-medieval contexts. They could have served many purposes, but Margeson suggests they may be associated with funerary clothing, as a number were found in association with burials at St Margaret's Church, Norwich, positioned along the arms of skeletons (1993, 20). Several were found during excavations in Southwark and in one case a small piece of textile preserved by waterlogged conditions was retained by a twist (Egan 2005, 62–4).

Six coins <149>, <154>, <181> and <235>–<237> were recovered from coffins or graves. A silver sixpence of Elizabeth I dated 1579 was recovered from coffin [1067]. A heavily worn and abraded hammered or struck coin was recovered from the fill of grave [998]. It is the size of a threepence and may be post-medieval. A George I silver half-crown (1714–27) also came from grave [998] and a George II copper-alloy halfpenny (1727–60) from the fill of coffin [762]. A much older coin, a copper-alloy farthing of James I (1602–25) also came from a coffin [1022]. The grave of Charles Shapley (d 1780) produced an undated, farthing-sized copper-alloy coin, in very poor condition. It may be that all of these finds are residual, but equally, they may form part of the burial practice of the time.

3.5 Clothing and personal items from the burials

Evidence for clothing included five copper-alloy buttons <141>–<145> from the grave of Richard Gideon Hand [622],

a member of the prosperous Hand family who ran the Chelsea Bun House in Pimlico (above, 3.1). They are c 18mm in diameter, plain with simple attachment loops on the reverse. Mr Hand may have been buried in one of his own jackets, an everyday item of clothing and not made for the funeral (Janaway 1993, 114). A white glass button <116> with a diameter of 9mm from the fill of grave of skeleton [323] may be from a shirt or other undergarment.

Two delicate mother of pearl or shell buttons <123> and <124> were recovered from the grave of skeleton [124]. These may come from clothing such as a cardigan or jacket, as recovered from the grave of Thomas Meacham at Christ Church Spitalfields (Molleson and Cox 1993, 203), which could have been worn under a shroud or alone, if the body was buried in everyday clothes.

It was not usual to include grave goods in 18th- or 19th-century burials, although at Chelsea Old Church there was the notable example of Dr Edward Chamberlayne (1616–1703), who was buried with the some of his books. He had instructed that the books were to be covered with wax, but this preservative measure was apparently unsuccessful for when his tomb was later opened they were found to have almost completely decayed (Faulkner 1829, i, 243–4). Another possible example was found during the excavation. This is a complete, heavily smoked pipe of type OS10 (1700–40), which appeared to have been deliberately placed in a grave [45] antedating that of Richard Butler (d 1732) (<245>, Fig 34). Complete clay pipes are sometimes recovered from graves of this period and are often taken as evidence for the grave diggers smoking while they worked, although in this instance the pipe may have been associated with one of the four burials stacked in the grave.

The dead were more commonly buried with items of personal adornment. A small ring <162> with a diameter of 11mm and a small attached loop was recovered from the Long family vault. It may well be the remains of an earring or other item of jewellery. It is in very good condition and a shiny gold colour, although tests on the metal suggest it is copper alloy.

Two copper-alloy mounts were recovered: a bar-mount from grave [279] and a book clasp with zigzag decoration with the skeleton in coffin [1067] (<26> and <239>, Fig 35). It is possible that they adorned items such as a belt or book, but it is more likely they are accidental inclusions, particularly as the bar-mount has medieval parallels (Egan and Pritchard 1991, 214, fig 134).

Some finds appear to have entered grave fills accidentally, either by inclusion at the time of burial or mixed up in the earth that backfilled the grave. The leg and foot of a cast copper-alloy vessel <195> such as a cauldron or skillet was found in the grave of skeleton [1076]. The remains of an iron whittle knife <114> with an ivory handle was recovered from the fill of the Long family tomb. The handle has a lozenge-shaped section and the end is similar to pistol handles (Margeson 1993, 122, no. 770). The tomb also produced a complete ceramic wig curler <166> stamped 'WB' with a crown on each end. Both items date to the 18th century.

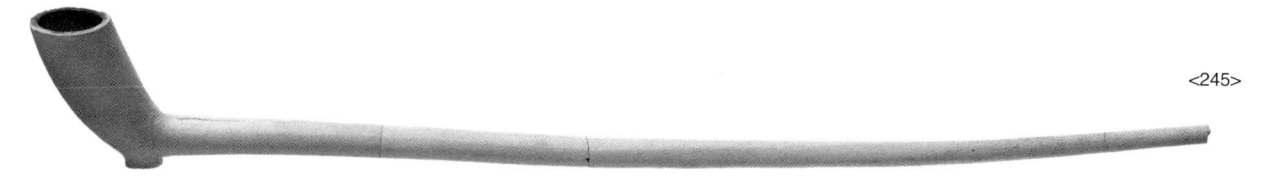

<245>

Fig 34   Complete clay tobacco pipe <245> made between c 1700 and 1732 (scale c 1:2)

<26>

<239>

Fig 35   Copper-alloy mount and clasp: a bar-mount <26>, probably a residual medieval artefact, from grave [279]; a book clasp <239> with fine linear decoration and a rivet hole, found with skeleton [1067] (scale c 2:1)

4

Life and death in Chelsea

Jelena Bekvalac and Tania Kausmally

4.1 Methodology

The skeletons recovered from the churchyard represented 290 individuals, of which 198 were recorded in detail on to the Wellcome Osteological Research Database (WORD) at the Museum of London; it is intended that this database will be made available online. The database contains all recorded metric data for each individual including information that, due to limited space, is not presented here. Most skeletons in the recorded sample were selected because they were relatively complete, although a subsample of 25 individuals for which there is biographical data (Table 2) were included in the database regardless of their completeness (Table 6). All data presented in this report are based on the recorded sample of 198 individuals.

The methodology was dictated by the WORD database following guidelines by Connell and Rauxloh (2003). Age at death of adults was based on morphological changes on the pubic symphysis (Brooks and Suchey 1990), the auricular surface (Lovejoy et al 1985) and the sternal portion of the ribs (Iscan et al 1984; 1985). For the subadults, age was based on the length of the long bone diaphyses (Maresh 1970; Scheuer et al 1980), development and eruption of teeth (Gustafson and Koch 1974; Smith 1991) and epiphyseal fusion (Scheuer and Black 2000). The age categories were divided into 12 stages (Table 7).

The sex of the adults was established from morphological features of the skull and pelvis (Phenice 1969; Ferembach et al 1980) and was scored on a five-point scale: male, possible male, intermediate, possible female and female (Buikstra and Uberlaker 1994). Sex determination for subadults was not attempted. Metric and non-metric data were recorded following guidelines by Brothwell (1981) and Buikstra and Uberlaker (1994). Tooth measurements were based on Buikstra and Uberlaker (1994). Height was estimated from the maximum femoral length using the formulae devised by Trotter and Gleser (1958; Trotter 1970). Pathology was recorded following Roberts and Connell (2004) with reference to Aufderheide and Rodriguez-Martin (1998), Ortner (2003), Roberts and Manchester (1995) and Salter (1999).

Data from Chelsea Old Church were compared with that from other excavated, contemporary cemeteries of varying social status in the London area (Table 8).

4.2 Preservation

Most (89%) bones in the assemblage were in good condition. However, many skeletons had suffered some truncation, mainly as the result of grave digging and building work in the churchyard, so that only a third were more than 70% complete (Fig 36). Skeletons in the subsample were between 2% and 97% complete, with almost half more than 50% complete.

Table 6 Details of the 25 individuals in the subsample identified through biographical information

Context no.	Name	Date of death	Known age	Osteological age	Osteological sex	Stature (cm)	Dental variations	Pathology	% of skeleton present
[35]	Gideon Richard Hand	1821	60 years	>46 years	male	169.9	transposition, rotation and impaction	osteoarthritis, ?gout, congenital, trauma (healed fracture)	95
[147]	Thomas Langfield	1808	67 years	>46 years	male	179.7	n/a	non-specific infection, osteoarthritis, trauma (healed fracture)	88
[171]	Christiana Hamilton	1748	neonate	-	-	n/a	n/a	none visible	6
[198]	Richard Butler	1732	44 years	36–45 years	male	169.1	n/a	ankylosis, metabolic (?healed rickets), enamel hypoplasia	65
[258]	Thomas Robson	17(3/5)1	-	26–35 years	?male	160.6	transposition and rotation	congenital, trauma (healed fracture), enamel hypoplasia	95
[430]	Martha Butler	1739	84 or 85 years	adult	female	n/a	n/a	HFI	7
[432]	John M[...]	-	-	>46 years	male	174.8	n/a	osteoarthritis, trauma (healed fracture)	82
[462]	Robert Butler	1712	61 years	36–45 years	male	170.6	n/a	trauma (healed fracture)	42
[525]	Charles Shapley	1780	70 years	>46 years	male	177.8	n/a	congenital, trauma (healed fracture), osteoarthritis, DISH, ankylosis	77
[622]	Richard Gideon Hand	1836	84 years	>46 years	male	176.9	n/a	metabolic (?osteoporosis)	35
[654]	Thomas Long	1827	66 years	>46 years	male	163.7	n/a	osteoarthritis, ankylosis, possible cyst in maxillary sinus	97
[681]	William Wood	1842	84 years	>46 years	male	170	n/a	DISH, osteoarthritis (diffuse), non-specific infection	75
[701]	Nicholas Adams	1827	78 years	>46 years	male	n/a	n/a	metabolic, congenital, ankylosis, erosive arthropathy	92
[713]	John Long	1822	68 years	>46 years	male	170	n/a	ankylosing spondylitis, osteoarthritis, non-specific infection, HFI	82
[719]	Harriet Long	1820	19 months	-	-	n/a	n/a	none visible	2
[722]	Catherine Long	1822	56 years	>46 years	female	159.8	transposition, rotation and impaction	osteoarthritis, sinusitis	90
[728]	Ann Wood	1807	53 years	adult	-	n/a	n/a	poor preservation, DJD in feet	18
[744]	John Long	1793	70 years	36–45 years	male	163.7	n/a	soft tissue trauma	39
[792]	Milborough Maxwell	1807	68 years	>46 years	female	157.8	n/a	osteoarthritis, non-specific infection	82
[976]	Edward Rainbows	1827	82 years	adult	-	169.5	n/a	Paget's disease, trauma (healed fracture)	33
[980]	Sarah Adams	1806	54 or 55 years	>46 years	female	162	rotation and crowding	non-specific infection, soft tissue trauma, DJD	51
[990]	Charity Adams	1781	32 years	26–35 years	?female	157.8	n/a	none visible	46
[1051]	?Collon/Cullum/Collins	?1797	-	-	-	n/a	n/a	none visible (truncated)	6
[1055]	? Mary ...od or ...ol	1735	?8- years	36–45 years	-	n/a	n/a	none visible (truncated); bulging occipital was noted	26
[1133]	Esther Long	1788	70 years	36–45 years	-	n/a	n/a	osteoarthritis	24

DISH: diffuse idiopathic skeletal hyperostosis
DJD: degenerative joint disease
HFI: hyperostosis frontalis interna

Table 7 Age categories (codes from Connell and Rauxloh 2003)

Code	Age group		Code	Age group
1	perinatal/neonate		7	18–25 years
2	1–6 months		8	26–35 years
3	7–11 months		9	36–45 years
4	1–5 years		10	>46 years
5	6–11 years		11	adult (indeterminate age)
6	12–17 years		12	subadult (indeterminate age)

Parish records and monumental inscriptions suggest that infants and young children are under-represented in the recorded sample. For example, monumental inscriptions suggest that 12 children were buried in the Adams family plot, but only four subadults were found there during the excavation. It is possible that the remains of most children mentioned in the inscriptions either had completely decayed or had been destroyed during subsequent burials, or were so ephemeral that they were missed.

Table 8 Excavated late 17th- to 19th-century cemeteries in London

Burial ground	Site address	Site code	Date	Status	Reference
Chelsea Old Church churchyard	2–4 Old Church Street, Chelsea, SW3	OCU00	18th–mid 19th century	medium/high	-
St Bride's crypt	St Bride's Church, Fleet Street, EC4	WFG62	1740–1852	high	Scheuer 1998
Christ Church Spitalfields crypt	Christ Church Spitalfields, Commercial Street, E1	CAS84	1729–1857	high	Molleson and Cox 1993
St Marylebone churchyard	St Marylebone School, Marylebone High Street, W1	MAL92 and MBH04	1750–1850	high	Powers 2005
St Pancras burial ground	Channel Tunnel Rail Link, St Pancras Terminus, York Way, N1	YKW01	1793–1854	high	Powers and White in prep
The New Churchyard	Broadgate, 123–229 Bishopsgate, EC2	LSS85	17th–18th century	mixed	White 1987
St Benet Sherehog churchyard	1 Poultry, EC2	ONE94	1670–1853	medium	White 2000
St Bride's lower churchyard	75–82 Farringdon Street, 20–30 St Bride Street, EC4	FAO90	1750–1849	medium/low	Conheeney and Waldron 2002
Cross Bones burial ground	Redcross Way, SE1	REW92	1729–1852	low	Brickley and Miles 1999

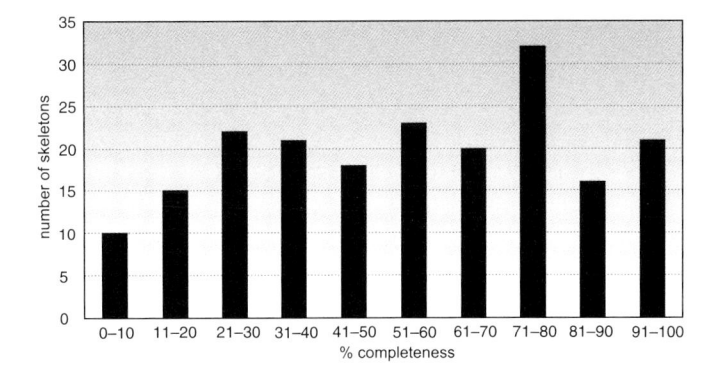

Fig 36 *Completeness of the skeletons in the recorded sample (n = 198) from 2–4 Old Church Street*

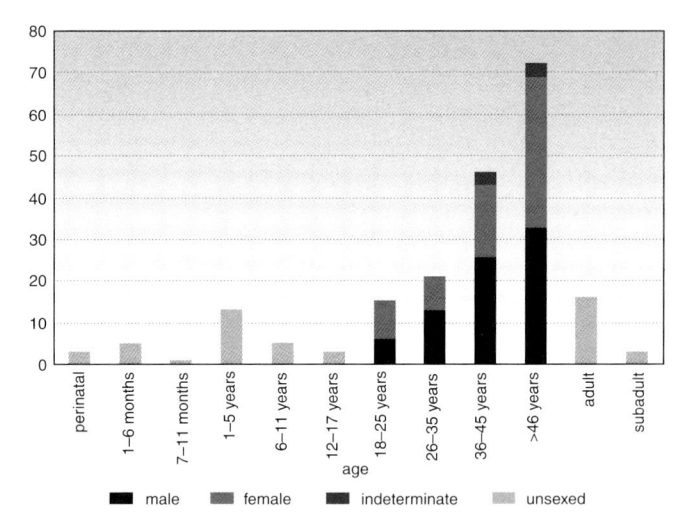

Fig 37 *Osteological age and sex distribution of the recorded sample (n = 198) from 2–4 Old Church Street*

Hair survived on the skulls of Nicholas Adams [701], Thomas Long [654] and three unidentified adult females. Both men had been buried in lead coffins in 1827. Preservation was particularly good in the case of Thomas Long, whose coffin had become waterlogged, so that hair, including possible traces of a beard, and finger and toe nails survived. Tracheal rings (throat region) were preserved in one male [453]; this is rare and, even if preserved, recovery is highly unlikely.

4.3 Population structure

Of the 198 individuals in the recorded sample 165 (83.3%) were adults and 33 (16.7%) were children or subadults (younger than 18) (Fig 37). The adults were almost equally divided between the sexes, comprising 78 males and 74 females. They were predominantly in the older category (>46 years old). Indeed, 17 of the 19 adults whose ages were known from coffin plates were aged between 53 and 84.

Females had a slightly higher death rate in the young adult category (18–25 years) whereas males were marginally higher in the middle-aged categories (26–35 and 36–45 years). There

was an almost equal number of males and females in the oldest age category (>46 years) (Fig 37).

The dangers of childbirth may be reflected in the higher prevalence of deaths in women in the younger age category (18–25 years) with a female:male sex ratio of 1.8:1. Indeed one woman in this age group [234] was found with a foetus of 38–40 weeks [2341] in the abdominal area. Another woman, aged 26–35 years [161], carried a foetus as young as 20–22 weeks [1611], possibly one of the youngest individuals ever recovered from a British archaeological assemblage. There is evidence to suggest that 32-year-old Charity Adams [990] may also have died shortly after giving birth.

Bills of Mortality in 18th- and 19th-century London (which did not include Chelsea) indicate that 50% of the population died before the age of 20. Before the 1760s those surviving into adulthood generally died between the ages of 20 and 60 years, but thereafter the average age of death appears to have gradually increased (Roberts and Cox 2003, 304). In the recorded sample at Chelsea Old Church 72 individuals (43.6%) died at an age over 46 years (Fig 38). It should be noted that osteological methods of establishing age-at-death tend to underestimate the ages of adults in this category. Indeed, biographical data indicates that several individuals in the subsample lived into their 70s and 80s (Table 6).

The subadults included 22 children under the age of 5 years and nine that were less than a year old. Coffin plates indicate that one baby lived only 5 days, Christiana Hamilton a matter of months, Matilda Long 2 months and Harriet Long 19 months. By 18th- and 19th-century standards the subadult mortality rate in the recorded sample from Chelsea Old Church was relatively low (16.7%). By comparison the subadult mortality rate at other excavated cemeteries of this period in Greater London ranges between 16.0% and 69.6% (Table 9). Generally, those cemeteries in poorer and/or more built-up areas, such as Cross Bones burial ground and St Bride's lower churchyard, have markedly higher mortality rates.

The Bills of Mortality for London during the 18th and 19th centuries suggest that about 40% of all deaths were of children of 5 years and under whilst the death rate between the ages of 6 and 20 years were as low as 6–7% (Roberts and Cox 2003, 304). While this is not entirely reflected by the assemblage from Chelsea Old Church it is clear that among the subadults in this group those below the age of 5 years are prevalent. A similar trend was found in most groups from excavated cemeteries of this period in Greater London (Table 9).

4.4 Accuracy of osteological age and sex determination

In order to test current ageing and sexing methods on the osteological data, the subsample was used as a 'blind' test. Sexing was tested on the morphological variations on the skull and pelvis. The latter produced the most accurate results. The techniques for establishing age-at-death, however, proved to be less accurate. Testing the auricular surface, pubic symphysis and the sternal rib ends revealed that the auricular surface was the least reliable, and there was a distinct tendency to underage in all three categories. The methods and results will be published in detail elsewhere (Bekvalac and Kausmally in prep).

4.5 Biometric data

Growth and development of subadults

Comparisons between dental eruption and diaphyseal long bone length may provide information on the rate of growth of subadults in a population. Age estimations of subadults are mainly based on studies of modern populations, and as such this should be considered when compared to archaeological assemblages (Maresh 1970; Gustafson and Koch 1974). Dental eruption, although still difficult to define clear-cut stages, is considered to be the more accurate method of ageing subadults osteologically (Hillson 1996, 138). Plotting the dental age against the diaphyseal long bone lengths, therefore, may provide an indication of the rate of growth in an individual.

It was only possible to make comparisons between 12 of the subadults. The limited sample size could not provide any statistically significant results (Fig 39). However, it did suggest that growth compared to dental development appeared to decrease with age. It is possible that the variation in age estimates is due to arrested growth due to childhood illness. This may be highlighted in individual [230] who suffered from rickets, which may partly

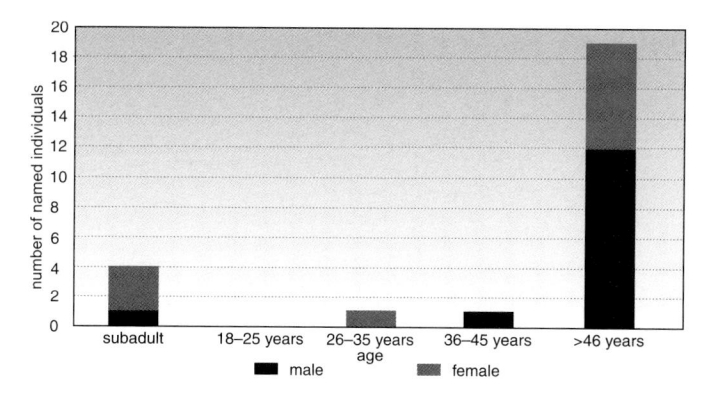

Fig 38 Osteological age and sex distribution of the named individuals (n = 25)

Table 9 Mortality rates of excavated groups from late 17th- to 19th-century cemeteries in London (for site status and references see Table 8)

	Total no. of individuals	Total no. of children	Children (%)	Children <5 years (%)	Children 6–17 years (%)	Total no. of adults	Adults (%)	Total no. not aged	Not aged (%)
Chelsea Old Church	198	33*	16.7	11.1	4.0	165	83.3	-	-
St Bride's crypt	131	21	16.0	10.7	5.3	110	84.0	-	-
Christ Church Spitalfields	968	215	22.2	-	-	660	68.2	93	9.6
St Marylebone	301	78*	25.9	22.3	3.3	223	74.1	-	-
St Pancras	715	183	25.6	19.7	5.9	532	74.4	-	-
The New Churchyard	388	122	31.4	15.7	15.7	266	68.6	-	-
St Benet Sherehog	230	65	28.3	18.7	9.6	165	71.7	-	-
St Bride's lower churchyard	533	193	36.2	-	-	340	63.8	-	-
Cross Bones burial ground	148	103	69.6	66.2	3.4	45	30.4	-	-

*: total includes unaged subadults

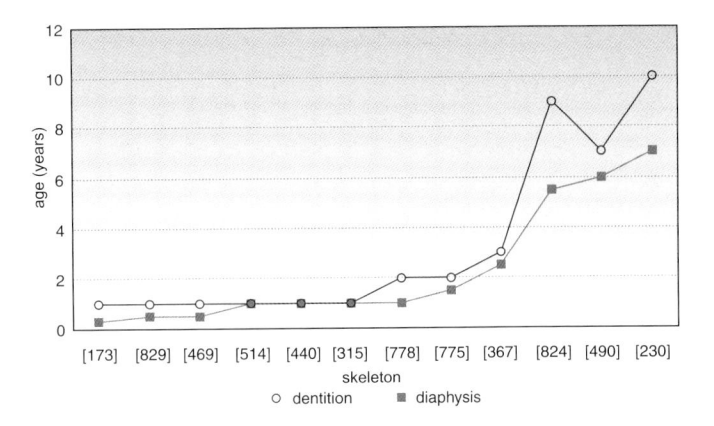

Fig 39 Comparison between dental age and age from diaphyseal long bone length for 12 subadults based on the length of the femora and humeri

explain the 3-year discrepancy in the age estimates. Major discrepancies do not appear until the age of 2 years, which may be linked to the age of weaning (Herring et al 1998). The limitations of the sample size, however, does not allow for any generalisation to be made about the Chelsea population.

Adult stature

Stature is thought to be a good indication of the health of a population. Growth of children is related to the availability of calorific foodstuffs which together with protein are essential for obtaining maximum stature during childhood (Roberts and Cox 2003, 308).

Females in the recorded sample were on average slightly taller than those found at other 18th- and 19th-century cemeteries in central London (Table 10). The males had a mean height of 170.3cm ± 3.94cm, similar to the average height of men buried at Christ Church Spitalfields and St Bride's lower churchyard, but noticeably shorter than the men at St Benet Sherehog (White 2000). However, they were on average taller than the men at Cross Bones burial ground (Brickley and Miles 1999).

It was possible to estimate the stature of 12 males in the subsample. The tallest at 179.7cm was Thomas Langfield [147], who was classed as a gentleman. The shortest at 160.6cm was Thomas Robson [258], whose teeth had enamel hypoplastic defects with marked bands extending to the roots (Fig 40). The hypoplasia may have been caused by prolonged periods of

Table 10 Stature (cm) of excavated groups from late 17th- to 19th-century cemeteries in London (for site status and references see Table 8)

	Males				Females			
	No.	Mean (cm)	Min (cm)	Max (cm)	No.	Mean (cm)	Min (cm)	Max (cm)
Chelsea Old Church	37	170.3	158.1	179.7	35	159.6	152.4	169.2
Christ Church Spitalfields	211	170.2	154.6	187.8	206	157.1	139.8	173.7
The New Churchyard	51	172.1	162.5	185.0	72	157.5	147.5	172.5
St Benet Sherehog	28	175.1	168.8	183.2	30	156.6	144.5	164.0
St Bride's lower churchyard	150	171.1	n/a	n/a	96	156.8	n/a	n/a
Cross Bones burial ground	16	168.5	153.0	180.0	19	158.2	142.0	172.0

Fig 40 Banding on root and enamel of teeth of Thomas Robson [258] (scale c 2:1)

illness or dietary deficiency during his childhood, which may have resulted in arrested growth. The stature of four females in the subsample were estimated. The tallest was Sarah Adams [980] at 162cm and the shortest was Milborough Maxwell [792] at 157.8cm. The stature of the subsample proved to be in a similar range to the recorded sample.

There was little variation in stature within family groups (Butler, Hand and Long families). This is perhaps not surprising for individuals with similar social and genetic background.

4.6 Non-metric variation

A total of 48 non-metric morphological variations in the skeletons were recorded and compared with data from other cemeteries. It is still unclear what these variations mean, although it is thought that some may be genetic whilst the frequency of other traits may be influenced by environmental factors (Brothwell 1981, 90). They are included in this study to allow comparison with other excavated groups, so that in time meaningful patterns may emerge. The traits in the subsample were not markedly different from those in the other individuals in the recorded sample.

Cranial variation

In the recorded sample the highest prevalence of non-metric traits were those of parietal foramen (46.4%) followed by mastoid foramen (42.0%) and posterior condylar canals (32.4%) (Table 11). At Christ Church Spitalfields the prevalence of parietal foramen (68.0%) and posterior condylar canal (64.0%) was similarly high (Molleson and Cox 1993, 35), but interestingly the most common non-metric trait at Spitalfields were those of accessory infraorbital foramina (85.0%), which had a significantly lower prevalence at Chelsea Old Church (5.6%).

Post-cranial variation

In the post-cranial skeletons at Chelsea Old Church (Table 12) the relatively high prevalence of calcaneal double facets (50.0%) is notable, which was over twice the prevalence (23.1%) recorded at St Benet Sherehog (White 2000). This trait was similarly prevalent in the subsample.

A number of post-cranial variations are believed to be activity related, this includes squatting facets of the tibiae and *os acromiale* of the scapulae. Repeated and habitual movement may promote a facet-like extension of the ankle joint, and the squatting facets of the tibiae have been associated with extreme dorsiflexion of the ankle joints. In a modern population this is around 2% as seen at Christ Church Spitalfields (Mays 1998, 118), whilst at Chelsea Old Church the incidence of lateral squatting facets was as high as 27.9%. The distribution between males and females in the Chelsea Old Church assemblage appeared to be relatively even with 28.0% and 32.0%. The medial squatting facets showed a male to female distribution of

Table 11 Cranial and mandibular non-metric traits in the recorded sample, with comparative data from Christ Church Spitalfields, Cross Bones burial ground and St Benet Sherehog (for site status and references see Table 8)

Cranial and mandibular non-metric traits	Chelsea Old Church churchyard %	Christ Church Spitalfields %	Cross Bones burial ground %	St Benet Sherehog %
Accessory infraorbital foramen	5.6	85.0	-	-
Asterionic bone	16.1	24.0	-	-
Bregmatic bone	1.2	2.0	-	3.9
Coronal wormians	-	12.0	13.3	1.3
Epipteric bone	4.8	7.0	-	-
Foramen of huschke	15.0	2.0	-	-
Inca bone	1.2	-	-	-
Lambdoid bone	9.8	34.0	6.6	6.5
Lambdoid wormians	14.6	-	17.8	14.9
Mastoid foramen	42.0	26.0	-	-
Metopism	12.4	9.0	13.3	16.9
Multiple mental foramen	2.4	0.6	-	-
Mylohyoid bridge	14.6	-	-	-
Parietal foramen	46.4	68.0	-	-
Parietal notch bone	6.3	27.0	-	-
Posterior condylar canal	32.4	64.0	-	-
Sagittal wormians	1.1	3.0	-	3.9
Squamo-parietal wormians	-	9.0	-	1.3
Supraorbital foramen	30.8	-	13.3	-
Supraorbital groove	21.3	14.0	-	-
Torus auditivi	1.5	-	-	-
Torus mandibularis	2.9	3.5	6.6	0.2
Torus maxillaris	5.3	15.0	6.6	0.1
Torus palatinus	15.4	1.0	-	0.3

Table 12 Post-cranial non-metric traits in the recorded sample with comparative data from St Benet Sherehog (for site status and reference see Table 8)

Post-cranial non-metric trait	Chelsea Old Church churchyard %	St Benet Sherehog %
Accessory sacral/iliac facets	-	-
Acetabular crease	13.7	0.9
Acromial articular facet	2.3	-
Allen's fossa	4.7	-
Atlas – double facet	8.1	13.0
Atlas – lateral bridge	6.2	-
Atlas – posterior bridge	16.2	-
Atlas – transverse foramen bipartite	0.7	-
Calcaneal facet absent	10.3	-
Calcaneal facet double	50.0	23.1
Hypotrochanteric fossa	9.6	-
Manubrio-corpal synostosis	11.3	-
Os acromiale	9.2	-
Patella – bipartite	-	-
Patella – vastus notch	12.8	2.8
Septal aperture	4.4	-
Sternal foramen	4.6	9.3
Supracondylar process	-	-
Talus – os trigonum	3.3	-
Talus – talar facet double	12.7	34.0
Talus squatting facet	16.0	2.0
Third trochanter	8.6	-
Tibia – lateral squatting facet	27.9	3.0
Tibia – medial squatting facet	9.4	-

11.0% to 8.5%.

It is unclear why these variations were present. The accessory infraorbital foramina are most likely to be related to different genetic clusters. The high prevalence of squatting facets in both males and females is less clear and may be genetic or activity related.

Dental variation

Genetic or environmental factors may be the cause of dental variations within a population, whilst crowding may occur if there is insufficient space to allow for eruption. Genetic traits include Carabelli's cusps, an accessory cusp of the first maxillary molar, which is common in Europeans (75–85% of individuals; Hillson 1996, 91). Another genetic trait is the 'enamel pearl', which is a small nodule of enamel on the root surface of the teeth (ibid, 98). These traits may be recorded in teeth in order to generate a pattern and may help establish population groupings.

A total of 19 of the 74 females (25.7%) and 10 of the 78 males (12.8%) had one or more dental anomalies (Table 13). Gideon Richard Hand [35] had a number of dental variations of the canines including transposition and impaction. Another middle-aged male [532] had extensive crowding in the mandible of the central and lateral incisors as well as the canines. Enamel pearls were present in both maxillary third molars in a young female [161], whilst Carabelli's cusps were noted in a middle-aged female [18] in the upper first molars.

Table 13 Frequency of dental anomalies in males (n=10) and females (n=19)

Dental anomaly	Males No. of teeth affected	Males No. of individuals	Females No. of teeth affected	Females No. of individuals
Crowding	13	3	26	12
Impacted	5	3	4	3
Rotation single	12	8	17	8
Rotation bilateral	2	1	4	2
Enamel pearls	0	0	2	1
Transposition	12	4	1	1
Carabelli's cusp	0	0	2	1
Total	**44**		**56**	

4.7 Palaeopathology

Introduction

In the recorded sample seven categories of disease were recognised, with joint diseases and trauma the most frequent (Fig 41). The diseases observed in the skeletal remains were not necessarily the cause of death, and only partly reflect those that would have affected the living population. The diseases manifested osteologically were usually chronic ones that affected individuals for a long time, whereas acute diseases such as smallpox rarely had time to cause changes in a skeleton. The London Bills of Mortality in 1775 showed the main causes of death during this period were from infectious diseases such as tuberculosis and smallpox (Cox 1996, 74).

Joint disease

Diseases of the joints are generally the most predominant disease identified in archaeological skeletal material. The most common of these is osteoarthritis, which in clinical terms may be identified as joint pain and diminished joint space but in skeletal material is characterised by both subtle and fully

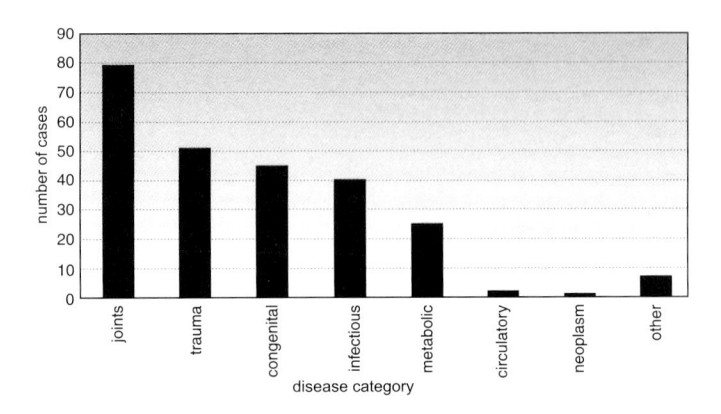

Fig 41 Disease groups represented in the recorded sample (some individuals are recorded within more than one disease category)

developed morphological changes in and around the joint (Ortner 2003, 545).

Osteoarthritis

The characteristic features of osteoarthritis in the recorded sample were deformation of the joint and eburnation of the articular surface. Osteoarthritis may be identified as either primary (age related) or secondary (pathologically related), affecting one or many joints (Table 14). All joint surfaces were examined for signs of degenerative joint disease (DJD), characterised in part as osteophytic lipping and pitting, which may be seen in some instances to be a precursor to osteoarthritis (Roberts and Manchester 1995, 103).

Osteoarthritis affected 45 adults (27.3%) in the recorded sample, similar to the proportion of affected individuals (26.3%) in the assemblage from Christ Church Spitalfields (Roberts and Cox 2003, 352). They included seven males and three females in the subsample, all of whom were over 60 years of age apart from Catherine Long [722], who was 56 years old. The others were all in the older age categories (36–45 years and >46 years) and consisted of 17 females, 14 males and four of indeterminate sex. Contemporaneous cemetery populations also show a higher prevalence of osteoarthritis with age, as correlated at Christ Church Spitalfields and modern populations, and for females to be slightly more affected (Waldron 1993, 67).

In both males and females the right side of the body was more affected by osteoarthritis. In the weight bearing joints the females, such as Esther Long [1133], appeared to be more affected in the knees, and the males in the hips. This pattern is also seen in studies on modern populations (Sisman and Hochberg 1993). Interestingly, the opposite was seen at St Benet Sherehog (Miles and White in prep). The hands and wrists in both sexes were also frequently affected, especially the thumb joint (Fig 42). In some individuals, such as Gideon Hand [35], Thomas Langfield [147] and Milborough Maxwell [792], both hands were affected.

A total of 31 individuals suffered osteoarthritis in more than one joint, including two older females [474] and [716] and an older male [593] with diffuse osteoarthritis involving many joints. William Wood [681] had diffuse osteoarthritis,

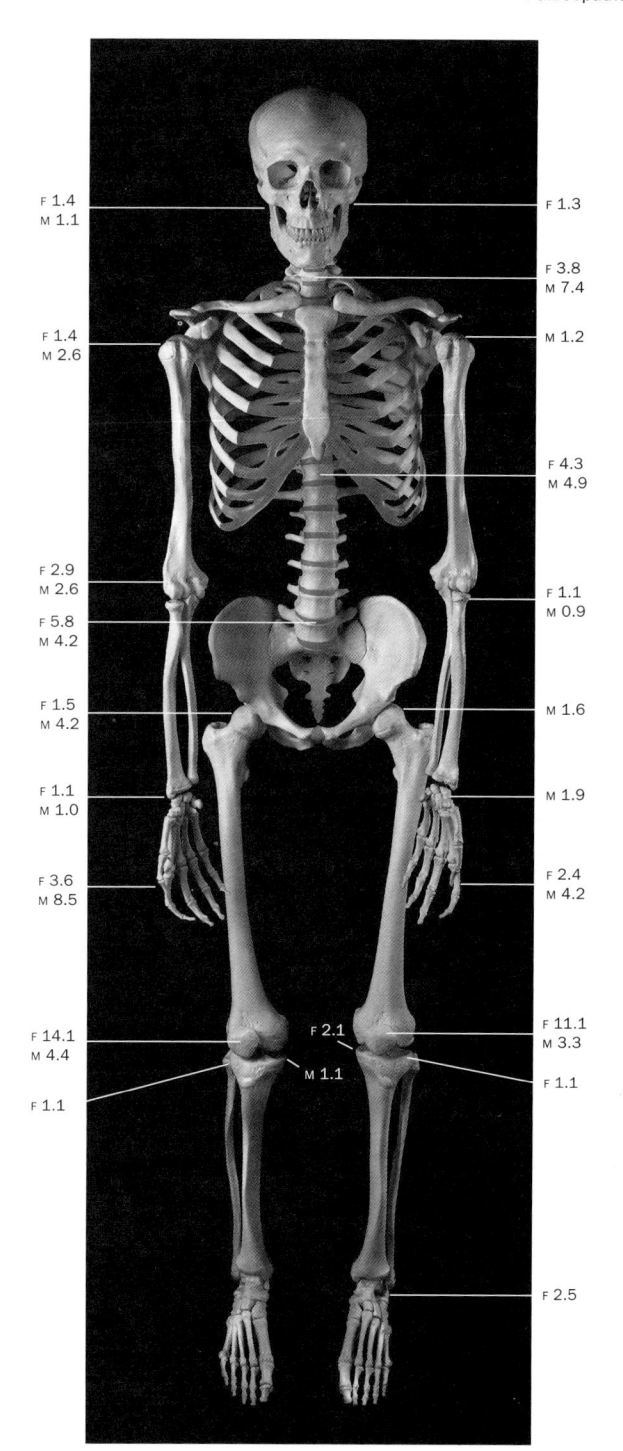

Fig 42 *Prevalence distribution of joints affected with osteoarthritis expressed as a percentage of the number of cases per number of males (M) and females (F) respectively in the recorded sample*

Table 14 *Number of individuals with joints affected with osteoarthritis in the recorded sample*

Joint affected	Total no. of individuals affected	Males	Females	Indeterminate sex
Shoulder	4	3	1	0
Elbow	8	5	3	0
Wrist/hand	33	16	13	4
Hip	6	4	1	1
Knee	12	3	9	0
Feet	11	7	3	1

particularly affecting the hips, that would no doubt have impeded movement. He also had diffuse idiopathic skeletal hyperostosis (DISH), a disease associated with diet and obesity, a factor to consider in relation to his osteoarthritis.

Osteoarthritis secondary to trauma was observed in three males, [323], [948] and Robert Butler [462], whose condition was probably due to a malaligned fracture of the left humerus. In all cases the secondary osteoarthritic changes were the result of an alteration in normal joint articulation.

Spinal joint disease

Spinal joint disease was observed and recorded as any changes affecting the integrity of the vertebrae. This may be expressed individually or collectively in fusion, osteoarthritis, osteophytic lipping, intervertebral disc disease (IVD) and Schmorl's nodes. The latter are observed as a depression in the main area of the vertebral body and linked to load bearing (Roberts and Manchester 1995, 107).

Spinal joint disease was found in 66 males, 50 females and four individuals of indeterminate sex. Degenerative changes of the cervical and lumbar vertebrae were more severe and prevalent in the females. Osteoarthritis was more common in the cervical vertebrae of males and the lumbar vertebrae of females (Fig 42), indicating a difference in the manipulation of the spine between the sexes and a variation in stresses in these vertebrae.

The thoracic and lumbar vertebrae of the males were more frequently affected with osteophytosis and Schmorl's nodes. Male thoracic vertebrae were nearly four times (35.7%) more affected with Schmorl's nodes than female ones (9.7%), suggesting perhaps that the males were more frequently engaged in activities involving load-bearing stress. Interestingly in contrast, the males and females at Cross Bones burial ground in Southwark were equally affected suggesting that both sexes were engaged in such stress-related activities (Brickley and Miles 1999, 38). This may reflect differences in lifestyles and working conditions between females living in slum conditions in Southwark and their comparatively well off counterparts in Chelsea.

The rates of intervertebral disc disease were relatively similar for the males and females with the cervical and lumbar vertebrae being most affected. This does not appear unusual for a population primarily of older individuals where intervertebral disc disease is associated with an increase in age.

Groups from other 18th- and 19th-century cemeteries also have a high frequency of spinal disease, which has become increasingly common over the period from the Middle Ages to modern times (Roberts and Cox 2003, 353). The variation in spinal disease and osteoarthritis of the spine is an important indicator of change of activity through time indicating, perhaps, differences in social status, physical labour versus bad posture (ibid), different roles of the sexes and an increase in age at death. Those cases identified in the group from Chelsea Old Church would seem to be more related to an older age group with a clear difference in the activities of males and females.

Ankylosing spondylitis

This disease results in the calcification of the spinal ligaments producing a rigid 'bamboo'-like quality to the spine (Aufderheide and Rodriguez-Martin 1998, 102). Its cause is unknown, but it appears to have a strong genetic factor.

Sixty-eight-year-old John Long [713] had changes characteristic of ankylosing spondylitis. His 66-year-old brother, Thomas [654], had unilateral sacroiliac fusion on the left side but did not show any other features associated with ankylosing spondylitis. However, there were no such changes in his father and mother, John [744] and Esther [1133].

Erosive arthropathy

Gout is one type of erosive arthropathy which is metabolic in origin with the increased levels of uric acid and deposition of urate crystals in the joints causing destructive lesions, usually affecting the big toe. Sufferers may have a genetic predisposition that is influenced by diet and the environment (Aufderheide and Rodriguez-Martin 1998, 108). One individual, 60-year-old Gideon Hand [35] manifested such lesions in his right big toe. As the proprietor of the Chelsea Bun House Gideon Hand would have had the social standing to over indulge in food and drink, which might have been a causal factor of his condition.

Ankylosis

Ankylosis as a term means fusion and this can occur throughout the elements of the skeleton and can be secondary to other diseases such as DISH, ankylosing spondylitis and traumatic arthritis (Aufderheide and Rodriguez-Martin 1998, 97, 102, 105).

The main joint affected in the recorded sample was the sacroiliac joint with 15 (7.6%) individuals exhibiting either unilateral (predominantly on the right side) or bilateral fusion. Four of these were from the subsample [198], [525], [654] and [701] and in the older age category apart from 44-year-old Richard Butler [198]. Of the remaining individuals affected with sacroiliac fusion nine were males and two females, again in the older age category. In Charles Shapley [525], two other males [453], [805] and one female [716] this could possibly be associated with DISH.

Ankylosis from possible trauma of the left tibiofibular joint was seen in an older male [188] and of the right ankle joint in an older female [918].

Diffuse idiopathic skeletal hyperostosis (DISH)

Diffuse idiopathic skeletal hyperostosis, which is generally found in males over 40, is characterised by fusion of the spine due to ligament ossification without intervertebral disc disease and has the appearance of 'dripping candle wax' (Rogers and Waldron 1995, 49). It may be linked to a rich diet, obesity and the occurrence of late onset diabetes (Roberts and Manchester 1995, 120).

Skeletal changes associated with DISH were observed in ten (5.1%) individuals in the older age category. Nine were males, including 70-year-old Charles Shapley [525] and 84-year-old William Wood [681], who both had the classic 'candle wax' fusion of the thoracic vertebrae. Documentary sources suggest that both of the named individuals could afford to eat and drink well, and William Wood's occupation as a butcher might also have provided an opportunity for over indulgence. Christ Church Spitalfields with 5.8% affected (Roberts and Cox 2003,

311) was equally high and suggests that both populations had a propensity to over indulge. By contrast, no cases of DISH were found at the low-status cemetery of Cross Bones burial ground in Southwark (Brickley and Miles 1999, 38).

Trauma

A total of 44 individuals (22.2%) exhibited evidence of trauma. This group comprised 30 males (68.2%), 12 females (27.3%), and two individuals of indeterminate sex (4.5%). None of the 33 children recorded were affected.

Fractures

The most frequent trauma recorded was fractures. A total of 24 individuals (12.1%) had sustained a total of 35 fractures (Table 15). Fractures may have been caused by external factors such as accidents or violence or be related to pathological conditions such as osteoporosis and osteomalacia. Whether an individual had sustained a fracture from an accident or violence is very difficult to determine, though inferences may be drawn from certain types of trauma.

Males were more frequently affected than females in all areas of the body apart from the spine. A total of 16 males (8.1%), six females (3.0%) and two of indeterminate sex (1.0%) had sustained fractures. Calculating the prevalence by number of each sex within the population 20.5% of all males and 8.1% of all females had fractures. The torso and the arms were the most frequently affected. Two females, [152] and [587], displayed classic Colle's fractures of the radii associated with osteoporosis.

Four individuals had multiple fractures affecting more than one bone. Three were middle-aged to elderly males [668], [948] and [782]. One was an elderly female [152] who most likely suffered multiple fractures as a consequence of osteoporosis.

Most of the injuries were simple closed fractures and the majority were well healed and displayed no sign of infection. A total of three fractures displayed malalignment: a right

clavicle [453], the left humerus of Robert Butler [462] and a right radius [668], which despite the malalignment were well healed.

The fracture rate was relatively high at Chelsea Old Church (12.1%) compared with recorded groups from other post-medieval cemeteries in London (Table 16). This was only exceeded by the low-status cemetery Cross Bones burial ground (30.4%) and the high-status cemetery of St Marylebone (16.9%).

Blunt force trauma

Blunt force trauma to the skull was noted in two adults: one elderly male [285] and one female [608]. The male had received a blow to the back of the head (occipital bone) and the female to the frontal bone. It was not possible to establish whether these were sustained by accident or an act of violence.

Rotator cuff injuries

Rotator cuff injuries were present in three elderly females [19], [39] and [697]. The rotator cuff muscles control rotation of the shoulder and may be exposed to strain, especially when throwing and the arm is above the head. A sudden sharp pain in the shoulder would indicate a possible rupture of a tendon, while a gradual onset is more likely to be inflammation. Age

Table 16 Number of individuals with healed fractures in the recorded sample and from late 17th- to 19th-century cemeteries in London (for site status and references see Table 8)

	Total no.	No. of individuals affected	% of individuals affected
Chelsea Old Church	198	24	12.1
Christ Church Spitalfields	968	51	5.3
St Marylebone	301	51	16.9
St Benet Sherehog	230	11	4.8
St Bride's lower churchyard	533	51	9.6
Cross Bones burial ground	148	45	30.4

Table 15 Number of individuals with fractures per body area calculated against total number of males (n = 78), females (n = 74) and indeterminate (n = 13) in the recorded sample (some individuals had more than one area affected)

Body area	Males with fractures (n = 16)	% of total males (n = 78)	Females with fractures (n = 6)	% of total females (n = 74)	Indeterminate (n = 2)	% of total indeterminate (n = 13)
Skull	2	2.6	-	-	-	-
Spine	2	2.6	-	-	-	-
Ribs	6	7.7	2	2.7	1	7.7
Arms	7	9.0	2	2.7	1	7.7
Hands	4	5.1	1	1.4	-	-
Pelvis	1	1.3	1	1.4	-	-
Legs	1	1.3	1	1.4	-	-
Feet	3	3.8	-	-	-	-
No. of individuals	16	20.5	6	8.1	2	15.4

degeneration of the tendon is a constant predisposing factor (Crawford Adams 2001, 230).

Soft tissue trauma

Occasionally muscle tissue will respond to trauma by producing bone directly in the muscle tissue itself, a condition known as *Mysoitis ossificans traumatica* (Ortner 2003, 134). In instances where this forms part of the existing bone tissue, bony protuberances may occur.

Evidence for soft tissue injury was evident in 12 individuals (6.1%). All were in the older category. The bones affected were the scapulae of one female [19] and two males [311] and [926], the femora of one female [436] and one male [948], the tibiae of two males [898] and [1157], the clavicle of a male [641], the occipital bone of a male [453], the fibula of a female [980] and the feet of two males [453], [466] and one female [1175].

Post-mortem examinations

Post-mortem examinations had been undertaken on two middle-aged males. One [805] had a craniotomy (removal of the skullcap). This individual had DISH, bilateral ankylosis of the sacroiliac joint as well as sinusitis; it is unclear whether the removal of the skull cap was carried out due to any of these conditions. The other [359] had his manubrium (breast bone) cut through, but exhibited no pathologies or other surgical intervention.

Such examinations were not uncommon in the 18th century with examples recovered from a number of cemeteries across London such as for example Christ Church Spitalfields (0.72%), Cross Bones burial ground (1.35%) and St Bride's lower churchyard (3.38%) (Roberts and Cox 2003, 315). The majority of the autopsies were restricted to the removal of the skullcap (Waldron 1993, 87).

Congenital disease

A total of 57 (28.8%) individuals from this sample had congenital malformations, most of which affected the spine (Fig 43).

Spina bifida occulta

Spina bifida is recognised as the failure of fusion of the posterior neural arch. It is most commonly seen in the sacrum and can involve one or more of the sacral vertebrae. In severe cases this may have a neurological impact, although the less debilitating form spina bifida occulta is more common in the archaeological record (Ortner 2003, 468).

Two individuals (1.0%) had spina bifida occulta affecting the sacrum, with the cleft occurring at the first sacral vertebra and continuing down to the fifth sacral vertebra. One was an older male [516] and the other a child [383] of approximately 11 years of age. The Christ Church Spitalfields population also had a low rate of spina bifida occulta at 2.3% (Waldron 1993, 87).

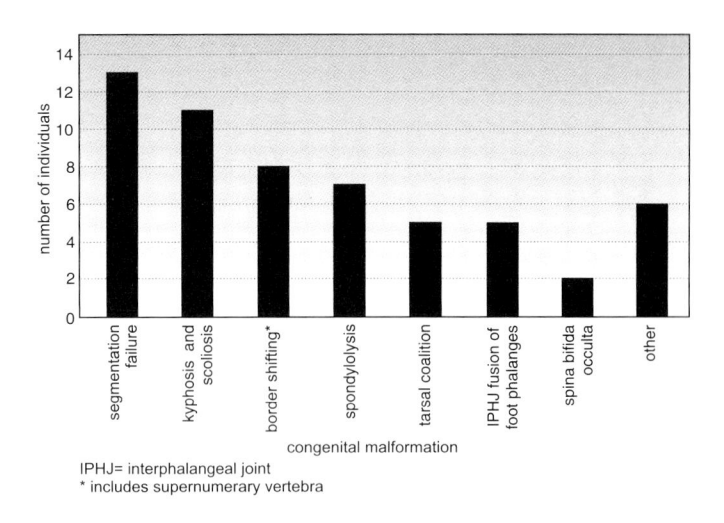

IPHJ= interphalangeal joint
* includes supernumerary vertebra

Fig 43 *Distribution of congenital malformations in the recorded sample*

Curvature of the spine

Congenital abnormal curvature of the spine was observed in 11 (5.6%) individuals with a varying degree of severity: five males [485], [593], [668], [739] and [951], five females [19], [104], [407], [434] and [1014] and one individual of indeterminate sex [780]. The cortical index results of the second metacarpal of one male [593] indicated osteopenia to be present which could relate to the curvature of the spine.

Slight scoliosis (lateral curvature with rotation) was seen in an older female [19] and a male [739] of 26–35 years with only slight changes in the lower thoracic and lumbar vertebrae. Kyphosis (angular deformity) was seen in two older males [485] and [668], the latter possibly as the result of trauma. The most severe malformations were in an unsexed adult [951] 26–35 years old and an older female [407]. The former had severe lateral curvature and wedging particularly of the eighth to 12th thoracic vertebrae, which also resulted in the flattening and deformity of the ribs. The older female had severe curvature to the spine, particularly in the thoracic vertebrae with gross distortion and fusion of the ribs, suggesting advanced kyphoscoliosis (Fig 44).

These deformities may have been severe enough to have affected the overall well being of these individuals. They also may have caused difficulty with mobility, eating and even breathing, particularly with the gross changes in the ribs that may have hindered respiration.

Spondylolysis

Another common malformation is spondylolysis, where a congenitally weak vertebra separates into two parts. This generally occurs at the fifth lumbar vertebra, possibly caused through recurrent stress of lifting with a bent back (Roberts and Manchester 1995, 78).

Seven individuals (3.5%), four males including Gideon Richard Hand [35], Nicholas Adams [701], [709] and [754], and three females [782], [888] and [1023], had spondylolysis.

Fig 44 *Severe congenital abnormality of the spine and ribs in an older female [407] (scale c 1:2)*

Linking such malformation to occupation is problematic, as the contributory causes are still not fully understood. Therefore any speculation associating the trade of the named individuals to this condition would be highly tentative. The rate of spondylolysis at Chelsea Old Church is similar to that found at Christ Church Spitalfields (2.8%), where also more males were affected (Waldron 1993, 87).

Nicholas Adams also had the most congenital malformations of all the individuals having spondylolysis, a supernumerary vertebra and a defect of the second cervical vertebra foramen. They do not appear to have affected him too untowardly as he married twice and lived to be 78 years old.

Tarsal coalition

Tarsal coalition was identified in five (2.5%) individuals [115], [161], [169], [363] and [918]. All were expressed in either unilateral or bilateral coalition of the navicular and calcaneus and interestingly all were female except for one adult [169] with no ageing or sex criteria present. The cause is unknown but it may be congenital or associated with secondary infection or trauma (Aufderheide and Rodriguez-Martin 1998, 75). Interphalangeal fusion of the foot phalanges was seen in five individuals [161], [258], [453], [525] and [898], two of them named, Thomas Robson and Charles Shapley.

Vertebral anomalies

Vertebral anomalies were observed as border shifts, segmentation failure other than spina bifida occulta (non-fusion of the neural arch) and developmental errors. Twenty-seven individuals expressed vertebral anomalies with two individuals expressing two types of anomalies.

Border defects can be expressed in numerous variations throughout the spine as described by Barnes (1994). Appearances of some of these defects are extra vertebrae (13th thoracic, sixth lumbar) and lumbarisation or sacralisation of the fifth or sixth lumbar vertebra. Individuals can express more than one anomaly and this was observed in the recorded sample. Six individuals (3.0%) had a supernumerary vertebra, including Nicholas Adams [701]. Sacralisation was observed in a young male [47] and a male 36–45 years [339], who also had an extra lumbar vertebra. Lumbarisation was identified in one young female [790]. The overall rate for border shifting was low with the predominant cause being a supernumerary lumbar vertebra.

Segmentation failure in the sacral region was the most commonly observed anomaly with 13 (6.6%) individuals affected. This failure was most often observed in the non-fusion of the neural arch of the first and third sacral vertebra. Six individuals had other various vertebral anomalies that were asymptomatic.

Infectious disease

Infectious diseases were rife during this time and included tuberculosis and syphilis. In this assemblage, perhaps surprisingly, syphilis was not found.

Tuberculosis

Tuberculosis is in essence a disease of urbanisation. It is an 'acute or chronic infection of soft or skeletal tissue by *Mycobacterium tuberculosis* or *M. bovis*' (Aufderheide and Rodriguez-Martin 1998, 118).

By the 17th century tuberculosis had reached epidemic proportions and was responsible for the highest rate of mortality in London after the plague. The Bills of Mortality indicate that it was responsible for 25% of deaths by the end of the 18th century (Roberts and Cox 2003, 338).

Only two individuals had bony skeletal lesions readily associated with tuberculosis. One was a young female of 18–25 years [754], the other a female of 36–45 years [339]. The tuberculous changes in both skeletons were in the spine, which is one of the most common sites for tuberculous lesions.

The stages of the disease can involve the formation of an abscess, perforation of the abscess, collapse of the vertebrae and bony fusion of the vertebrae (Roberts and Manchester 1995, 138). In both cases destruction was seen of the discs, and in individual [339] there was evidence for possible abscess formation with calcification of the pus. Neither of the individuals had bony fusion of the vertebrae.

A male of 36–45 years [836] with changes identified on the superior body of the first sacral vertebra were similar to those seen in tuberculosis. Another explanation for these changes was a herniated disc and that the destruction was due to the rupturing and damage between the discs.

The fact that only two cases of tuberculosis were found with bony lesions may not be quite so surprising as first thought. Such a low rate may also be seen when compared to other sites such as St Marylebone (Powers 2005, 10) with four cases, Christ Church Spitalfields with two cases and St Bride's lower churchyard with four cases (Roberts and Cox 2003, 339). Although it was a prevalent disease of the time with a high mortality rate various factors should be considered. The sample of individuals excavated was just that and they may not be indicative of the true mortality rate from tuberculosis for the parish population. The prevalence of tuberculosis may have been higher but if its contagion and rapidity were such, then its affect upon the bones would be limited. Other factors such as housing conditions, nutrition and environment must also be considered. The parish in comparison to some of those within London was still relatively rural, not densely built or over populated and had large areas of market gardens, which may in part explain the low tuberculosis rate.

Periosteal bone reactions

A periosteal bone infection is generally regarded as a non-specific surface inflammation or periosteal reaction of the bones. 'The inflammatory process is manifest as fine pitting, longitudinal striation and, sequentially, plaque-like new bone formation on the original cortical surface' (Roberts and Manchester 1995, 129–30).

A total of 38 individuals (19.2%) in the recorded sample showed evidence of a periosteal bone reaction. The most commonly affected bones with periosteal infection were those of the legs, with the tibiae being the most markedly affected followed by the femur and fibula. The presence of these active and healed periosteal reactions might be seen as an indication of minor infections, possibly associated with stress, minor trauma or varicose vein problems.

One female [885] aged 26–35 years had changes in the frontal and parietal bones that may have been a meningeal reaction. Fifty-six-year-old Catherine Long [722] and a male [805] aged 36–45 years had maxillary sinusitis, which can be associated with a number of irritants including allergies, smoke and environmental pollution. A further seven (3.5%), including Thomas Langfield [147], had lesions on the visceral surface of one or more ribs, possibly the result of chest infections, pneumonia or even tuberculosis. Studies have shown that rib lesions may sometimes be correlated with tuberculosis, although such lesions are not necessarily indicative of this disease (Santos and Roberts 2001, 38–49; Lambert 2002, 281–92; Mays et al 2002, 27–86).

Metabolic disease

A number of metabolic disorders may be recognised in the skeleton. Among these are rickets, osteomalacia (adult rickets), osteoporosis and scurvy. In the 17th and 18th centuries these diseases were far from uncommon (Roberts and Cox 2003, 308) and have been observed in contemporary archaeological assemblages across London (Table 17).

Rickets

This childhood disease has three main causes: vitamin D deficiency, chronic renal insufficiency and renal tubular deficiency. It results in bones becoming deformed due to inadequate calcification and may cause bowing of the weight bearing limbs. Historical records show that rickets was rife in the 18th and 19th centuries. For example it was reported at Great Ormond Street Hospital in 1867 that as many as 33% of admitted children were rachitic (Wohl 1983, 56).

Evidence of rickets was found in two children in the recorded sample. One was a 1-year-old, [456], with signs of active rickets, who exhibited enlargement around the epiphyseal plates of the long bones and the costochondral junction (Fig 45). The other, [230], was an 11-year-old with healed rickets, who also had severe enamel hypoplastic defects, suggesting extended episodes of stress during childhood (Fig 46). Residual rickets was also apparent in five males [43], [1071], [856], [701] and [258] and four females [104], [957], [918] and [910]. Though the rickets had long healed

Table 17 *Prevalence of metabolic disease in the recorded sample and excavated groups from late 17th- to 19th-century cemeteries in London (for site status and references see Table 8)*

	Total no. of individuals	Osteoporosis		Osteomalacia		Rickets		Cribra orbitalia		Scurvy	
		No.	%	No.	%	No.	%	No.	%	No.	%
Chelsea Old Church	198	10	5.1	2	1.0**	11	5.6	18	9.1	1	0.5
Christ Church Spitalfields	968*	10	1.0	0	-	35*	3.6	141*	14.6	0	-
St Marylebone	301	0	-	0	-	11	3.7	13	4.3	4	1.3
The New Churchyard	388*	0	-	2	0.5**	14*	3.6	27	7.0	0	-
St Benet Sherehog	230	0	-	0	-	9	3.9	28	12.2	0	-
St Bride's lower churchyard	533*	8	1.5	0	-	27*	5.1	14*	2.6	0	-
Cross Bones burial ground	148*	0	-	0	-	10*	6.8	6*	4.1	0	-

*: source is Roberts and Cox (2003); note that the calculated prevalence rates for these are based on total number of excavated skeletons, not just those where pathology would have been observable. Where percentage is not provided the same approach has been adopted

**: based on total number of skeletons (all age categories)

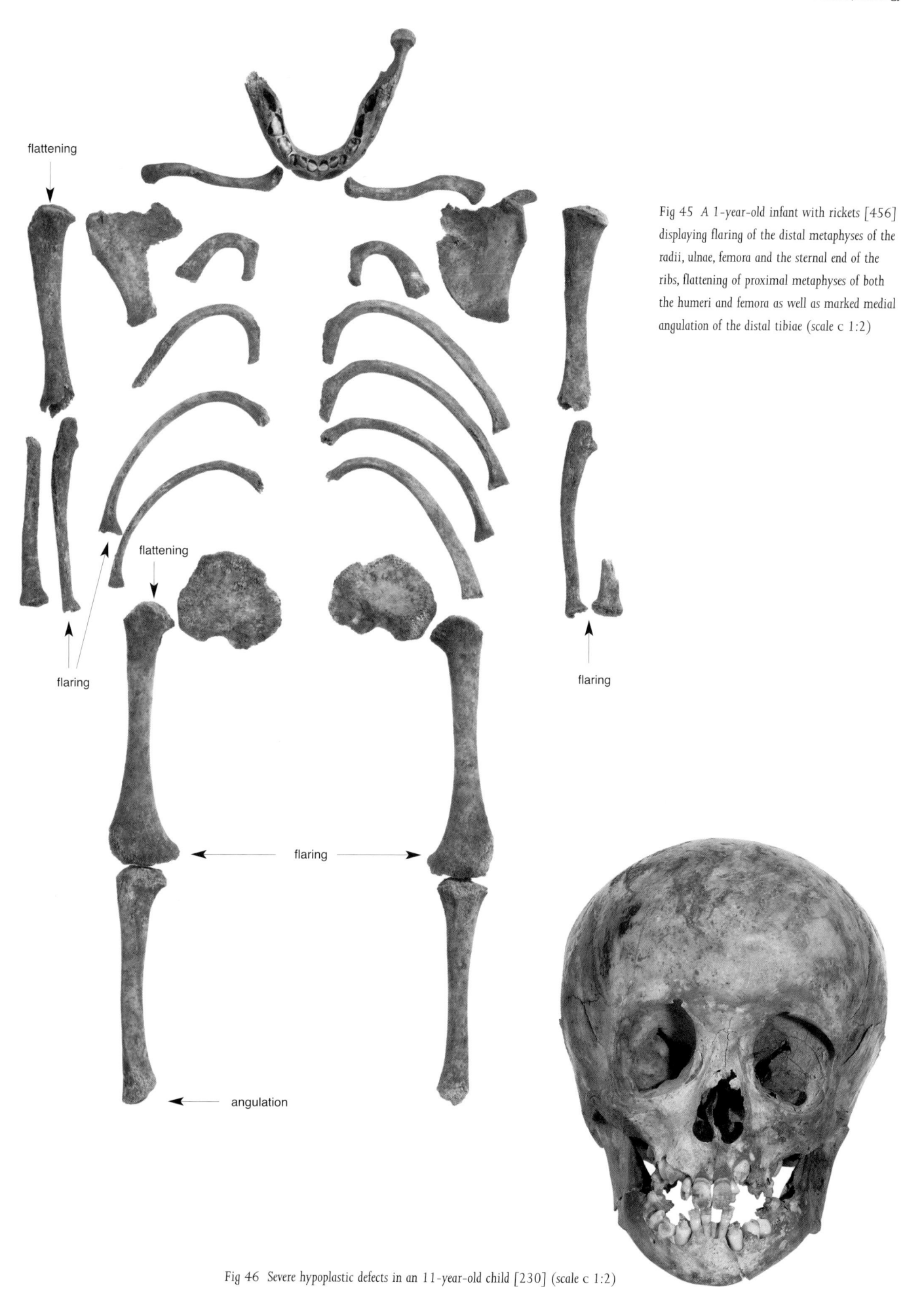

flattening

flattening

flaring

flaring

flaring

flaring

angulation

Fig 45 *A 1-year-old infant with rickets [456] displaying flaring of the distal metaphyses of the radii, ulnae, femora and the sternal end of the ribs, flattening of proximal metaphyses of both the humeri and femora as well as marked medial angulation of the distal tibiae (scale c 1:2)*

Fig 46 *Severe hypoplastic defects in an 11-year-old child [230] (scale c 1:2)*

in these individuals they exhibited the distinct bowing of the lower limb bones signifying vitamin D deficiency rickets during childhood.

Interestingly, the prevalence of rickets at 5.6% at Chelsea Old Church was comparable to that from cemeteries in poorer parts of 18th- and 19th-century London where there was evidence of the disease: St Bride's lower churchyard 5.1% (Conheeney and Waldron 2002), Cross Bones burial ground 6.8% (Brickley and Miles 1999) (Table 17). Contemporaneous cemeteries in prosperous parts of London appeared to have a lower prevalence rate: St Benet Sherehog 3.9% (White 2000), the New Churchyard 3.6% (White 1987) (Table 17).

The somewhat higher rate of rickets at Chelsea Old Church compared to sites of similar socio-economic status is puzzling. It may be a result of differential treatment during childhood or at an osteological level due to diagnostic disparity.

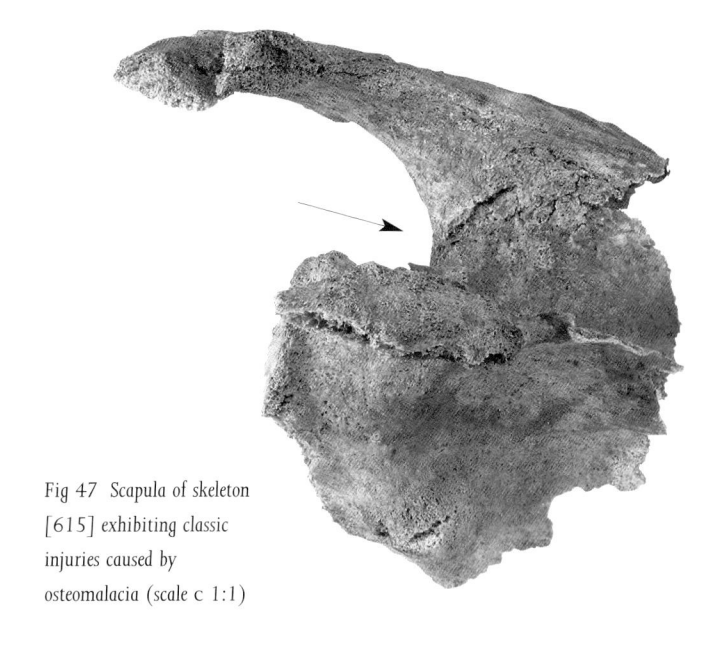

Fig 47 *Scapula of skeleton [615] exhibiting classic injuries caused by osteomalacia (scale c 1:1)*

Osteomalacia

Osteomalacia is a form of osteopenia and means 'soft bones'. It is a disease of adulthood, where calcium salts are not deposited in newly formed organic bone matrix. In advanced cases bones may soften and bend and become progressively deformed. In modern populations it has been seen in association with anorexia, weight loss, muscle weakness and widespread bone pain as well as bone tenderness and progressive bone deformity (Salter 1999, 187).

There were two clear cases of osteomalacia, [615] and [764], in the Chelsea Old Church assemblage (1.0%). Both were poorly preserved and neither could be aged or sexed. Both had multiple unhealed fractures in close proximity to the articular head of the ribs and fine hairline fractures on the neck of the acromion of the scapula (Fig 47, denoted by arrow).

Comparative data is uncommon which may be due to the often very poorly preserved remains in the archaeological records. Of all the comparative sites only two cases were recorded at the New Churchyard indicating a prevalence of 0.5% (White 1987, 11). New methods of diagnosis recently established (Ives 2005, 8), however, have improved the likelihood of diagnosis.

Osteoporosis

Osteoporosis is characterised by a decrease in bone amount, which increases skeletal fragility and the risk of fracture. There may also be vertebral collapse and extreme thinning of the flat bones in areas such as the iliac fossa in the pelvis. It is extremely common today and occurs in as many as one in four women and one in eight men over the age of 50. The risk factors are increased age, menopause, amenorrhea, insufficient calcium in the diet, eating disorders, smoking, excessive use of caffeine or alcohol and inadequate physical exercise (Salter 1999, 190).

In the archaeological record osteoporosis is found at a much lower rate than in the modern population. It is possible that it was less common in the past or that it is poorly detected

in the archaeological record. It is difficult to identify osteoporosis macroscopically and often X-rays are required to confirm its diagnosis (Brickley and Agarwal 2003).

Prior to examination of the cortical index a total of ten individuals were thought to have osteoporosis. Eight were elderly females, one male with possible osteoporosis [593] and one elderly individual of indeterminate sex [615]. Four of these had spinal collapse [407], [434], [593] and [1014] and in the remaining individuals the friability of the bones was the diagnostic criteria. The second metacarpal of seven of these individuals was examined or X-rayed (this bone was absent in the remaining three [152], [615], [1014]). Three showed little or no sign of osteopenia, but four showed definite thinning of the cortical index suggesting that some form or degree of osteopenia was present (Table 18; Ives in prep). The mean cortical index for young females is 50.9% and for young males is 44.4%. For older females the cortical index is 45.4% and for older males 43.3%. There are decreases in cortical bone amount in both sexes with increased age, with a greater decrease in the females. In an unusual trend, the males appear to have less cortical bone than the females and investigation of these results is ongoing.

Again the prevalence at Chelsea Old Church (5.1%) appears to be much higher than that of Christ Church Spitalfields (1.0%) and St Bride's lower churchyard (1.5%) (Waldron 1993, 80; Conheeney and Waldron 2002), possibly due to the refinement of diagnostic methods.

Cribra orbitalia

Cribra orbitalia is not associated with any one specific disease but is simply a description of porotic changes occurring in the orbital roof. The tendency in the past has been to associate it with anaemia (Roberts and Manchester 1995, 165), though this is not necessarily the case it does appear to be attributable to a metabolic disorder (Stuart-Macadam 1989, 212).

For Chelsea Old Church the prevalence was 9.1% falling

Table 18 Cortical index (CI%) of the second metacarpals of individuals who appeared to have macroscopic osteoporosis in the recorded sample (information courtesy of Rachel Ives)

Skeleton no.	Sex	Age (years)	Cortical index results (%)	Comments
[407]	female	>46	46.62	no osteopenia
[434]	female	36–45	50.07	no osteopenia
[483]	female	>46	29.07	very low CI, risk of osteoporosis
[587]	female	>46	42.61	no osteopenia
[593]	male	>46	38.11	quite low CI, likely osteopenia present
[628]	female	>46	40.34	quite low CI, likely osteopenia present
[892]	?female	>46	41.51	quite low CI, likely osteopenia present

below the average at Christ Church Spitalfields (14.6%; Molleson and Cox 1993, 43) and St Benet Sherehog (12.2%; White 2000), but still relatively high compared to other London sites (Table 17). Only around 2% of the Chelsea Old Church population was recorded as having more severe changes in the orbital roof.

Scurvy

Scurvy is caused by vitamin C deficiency, and is characterised by the lack of osteoblastic formation of bone accompanied by subperiosteal and submucous hemorrhages. The disease is more readily seen in children between the ages of 6 months and 1 year. The clinical features are an undernourished appearance, irritability, swelling and severe pain of limbs. The skin will be very tender over the affected bone (Salter 1999, 188).

A 1-year-old infant [440] showed signs of possible healed scurvy. The left radius and ulna exhibited swelling at the distal metaphyseal ends as well as flaring of the sternal rib ends. Macroporosity was present on the maxillary region, although no changes were noted on the sphenoid (a bone in the skull). Although these features may be caused by scurvy (Brickley and Ives 2006) it is possible that they may be related to rickets or another metabolic disease. Of the sites used for comparison in this study cases of scurvy were only recorded at St Marylebone (Table 17).

Circulatory disease

Osteochondritis dissecans is a disturbance in the normal blood circulation to the site of the skeletal pathology (Ortner 2003, 343). One possible case was noted in the Chelsea Old Church assemblage, an elderly male [593] exhibiting two smooth rounded lesions on the patellar joint surfaces.

Neoplastic disease

'Neoplasms or new growths are in essence the uncontrolled growth of tissue cells' (Roberts and Manchester 1995, 186). Neoplasms can be divided into two groups, those that are benign, the most frequently found, and malignant. The most common of the benign neoplasms in the skeletal record is the

'button' osteoma, generally found on the outer table of the skull (Roberts and Manchester 1995, 188). One was recorded on the frontal bone of an elderly female [446], being the only neoplasm in the assemblage.

Other palaeopathology

Hyperostosis frontalis interna (HFI)

Hyperostosis frontalis interna is located in the frontal bone of the skull, where thickening occurs in the inner surface, first appearing as small nodules and in later stages as a thickened undulating surface. The condition, which mainly affects elderly women, has been associated with obesity and diabetes mellitus (type II diabetes) (Nakhoda and Greene 2005).

Out of a possible 114 individuals where the frontal bone was present and able to be examined 6 displayed HFI. One was 68-year-old John Long [713], while the others were elderly females, including Martha Butler [430]. It is possible that in John Long's case the bony outgrowth on the internal surface of his skull was linked to other bony formations associated with his ankylosing spondylitis (Aufderheide and Rodriguez-Martin 1998, 102).

Paget's disease

Paget's disease is characterised by the gradual enlargement and deformity of one or more bones (Ortner 2003, 436). In modern populations the condition may affect as many as 4% of individuals over 55 years of age and is more common in men. The milder form may have very little affect on the diseased individual, but the more severe form may cause bone pain and the legs may become progressively bowed and the skull enlarged (Salter 1999, 199).

The skeletal remains of 82-year-old Edward Rainbows [976] showed the classic changes of Paget's disease (Fig 48). Although the skeleton was poorly preserved (33% survived) the skull, femora and tibiae were available for X-ray, which revealed changes to the skull and the femora. The tibiae, however, appeared completely normal (Fig 48).

Its occurrence at Chelsea Old Church was low at 0.6% of adults, though the figure is consistent with other post-medieval

a

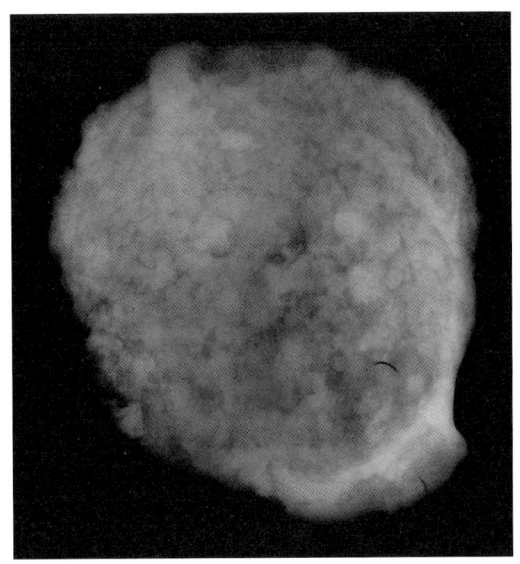

b

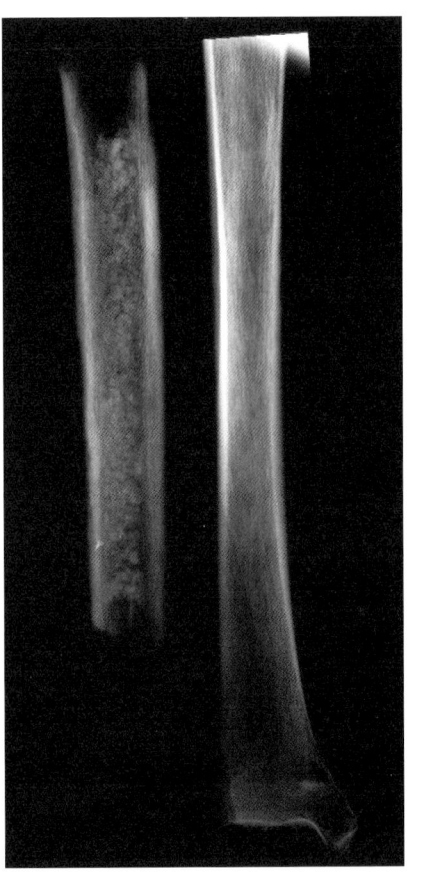

c

Fig 48 The skull (a) of Edward Rainbows [976] displaying changes caused by Paget's disease (scale c 1:1); bone thickening and irregular islands of dense new bone produce a characteristic mottled effect seen in the X-rays of Edward Rainbows's skull (b) and femur (c, left) whilst the tibia (c, right) is unaffected (scale b and c c 1:2)

cemeteries such as Christ Church Spitalfields (1.8%), Cross Bones burial ground (0.7%) and St Bride's lower churchyard (2.1%) (Roberts and Cox 2003, 355), all falling below the expected prevalence.

4.8 Dental pathology

There were 46 males and 42 females with observable teeth, including those in the subsample (Table 19; Table 20).

Dental caries (tooth decay)

Caries were present in 14.2% of the maxillary teeth. They occurred most frequently in right premolars, and were least common in right canines. Caries were less common in mandibular teeth, of which 8.5% were affected. The teeth most affected were the right second molars, while there were no caries observed in the mandibular central incisors, the lowest rate was in the left mandibular incisor (Table 19). Tooth decay

Table 19 Distribution of dental pathology in individual teeth (all individuals with permanent teeth in the recorded sample)

			Right										Left					
Maxilla	M3	M2	M1	P4	P3	C	I2	I1	I1	I2	C	P3	P4	M1	M2	M3	Total
Tooth positions available	49	72	74	76	76	81	81	87	81	81	80	77	76	72	69	53	1185
% lost post-mortem	12.2	6.9	12.2	11.8	7.9	13.6	23.5	20.6	22.2	21.0	11.3	10.4	10.5	5.6	4.3	7.5	13.0%
% lost ante-mortem	32.6	38.9	47.2	43.4	28.9	23.5	28.4	25.3	24.7	29.6	30.0	33.8	40.7	51.4	44.9	30.1	34.3%
% tooth unerupted	10.2	2.8	1.4	2.6	2.6	-	-	1.1	1.2	1.2	-	1.3	1.3	1.4	2.9	9.4	2.1%
% periapical lesions	2.3	2.9	4.1	2.7	2.7	1.3	-	-	-	3.8	3.8	1.3	2.7	2.8	3.0	-	2.1%
% periodontal disease	38.9	54.8	57.7	48.1	53.0	41.7	20.0	40.0	28.6	28.6	26.8	55.6	36.7	50.0	41.4	50.0	41.6%
Teeth available	24	42	36	35	49	56	50	48	45	42	52	46	38	34	35	35	667
% carious lesions	23.0	17.5	20.0	26.5	21.3	3.8	4.9	8.5	8.9	4.9	17.0	17.4	18.9	12.9	13.9	17.1	14.2%
% calculus deposit	80.0	87.1	73.3	67.6	67.4	5.6	48.9	57.4	54.5	46.1	60.9	73.3	77.8	80.0	81.8	68.6	62.3%
% enamel hypoplasia	-	-	7.7	3.1	11.1	37.3	21.1	28.3	20.5	17.5	36.9	11.6	8.3	7.1	3.1	-	15.3%

			Right										Left					
Mandible	M3	M2	M1	P4	P3	C	I2	I1	I1	I2	C	P3	P4	M1	M2	M3	Total
Tooth positions available	67	96	93	92	93	94	93	93	96	91	95	94	92	94	94	67	1444
% lost post-mortem	3.0	5.2	3.2	9.8	9.7	9.6	7.5	19.4	18.8	15.4	10.5	3.2	9.8	6.4	4.3	7.5	9.1%
% lost ante-mortem	41.8	52.1	62.3	33.7	25.8	13.9	22.5	26.9	27.1	19.8	14.7	27.7	36.9	57.4	52.1	38.8	34.4%
% tooth unerupted	7.4	2.1	3.2	1.1	1.1	1.1	1.1	-	1.1	1.1	1.1	2.1	1.1	2.1	2.1	7.5	2.0%
% periapical lesions	-	2.2	2.2	2.2	1.1	1.1	-	-	-	-	1.1	2.2	2.2	2.2	-	1.6	1.1%
% periodontal disease	36.7	51.4	59.3	56.5	57.7	51.7	57.4	48.8	43.9	43.7	42.8	52.7	55.6	50.0	51.3	41.3	50.4%
Teeth available	35	43	34	54	62	77	65	53	52	61	73	66	51	35	41	32	834
% carious	17.6	29.3	16.1	7.5	6.6	6.7	3.1	-	-	1.7	4.3	4.8	11.8	18.8	17.5	18.8	8.5%
% calculus deposit	70.6	73.2	78.6	74.5	71.7	71.6	83.1	90.4	82.7	83.3	77.1	71.4	79.2	74.2	25.0	77.4	74.8%
% enamel hypoplasia	-	-	7.1	3.8	19.2	4.2	27.4	26.5	3.3	2.6	9.7	5.9	11.5	38.5	20.7	30.6	18.2%

Dentition codes: I = incisor; C = canine; P = premolar; M = molar

Table 20 Distribution of dental pathology in males and females (excluding individuals of indeterminate sex) in the recorded sample

	Caries		Calculus		Hypoplasia		Periodontal disease		Periapical lesions		Tooth loss (before death)	
	male	female	male	female	male	female	male	female	male	female	male	female
Observable teeth	753	679	734	651	705	630	622	512	1339	1132	1067	910
Pathological teeth	91	93	551	464	136	96	344	202	15	26	376	306
% pathological teeth	12.1	13.7	75.1	71.3	19.3	15.2	55.3	39.5	1.1	2.3	35.2	33.6
No. of individuals with observable teeth	46	42	46	42	46	42	46	42	46	42	46	42
No. of individuals with dental pathology	26	29	45	39	25	22	37	29	11	14	42	39
% of individuals with dental pathology	56.5	69.0	97.8	92.9	54.3	52.4	80.4	69.0	23.9	33.3	91.3	92.9

affected 69.0% of the females and 56.5% of the males (Table 20).

Dental calculus

There are two forms of calculus (mineralised plaque): supragingival and subgingival, which accumulate respectively on the crowns and roots of teeth. In the permanent maxillary teeth the highest values for calculus were seen in the molars and the lowest in the anterior teeth (Table 19).

For the permanent mandibular teeth 74.8% were affected

with calculus deposits. The most affected teeth were the incisors, a much higher rate than was seen in the maxillary incisors (Table 19). The concentration of calculus deposits was greater on the mandibular teeth overall compared to the maxillary. The males at 97.8% were more affected than the females at 92.9% (Table 20).

Enamel hypoplasia

Enamel hypoplasia is 'a deficiency of enamel thickness, disrupting the contour of the crown surface, initiated during enamel matrix secretion', which may be multifactorial and related to illness, trauma, stress, nutrition and environmental factors (Hillson 1996, 165–6).

In general males appeared to be more affected. Of the teeth present among the males 19.3% displayed enamel hypoplasia. Of all the teeth observed and affected 46.0% had mild changes, 38.9% medium and 15.1% severe. Two males with particularly marked hypoplasia were Richard Butler [198] and Thomas Robson [258]. In females 15.2% of teeth were affected to some degree with enamel hypoplasia. Of all the teeth affected 48.9% had mild changes, 42.4% medium and 8.7% severe. The canines were the most frequently affected teeth in both sexes showing the most marked changes in the middle crown area. In females the upper crown area of the central incisors had the most severe changes whereas in males it was the upper first molars.

Periodontal disease

Evidence for periodontal disease (alveolar resorption) was found in 37 males and 29 females (Table 20). It has been suggested that periodontal disease is linked to age (Hillson 1996, 266), which is perhaps relevant for this assemblage as the majority were older individuals. The males also exhibited the most severe periodontitis (Grade 3) at 16.4% compared to females at 8.0%. The highest rates of peridontitis were seen in the right maxillary and mandibular first molars (Table 19).

Abscesses (periapical lesions)

A periapical lesion may be identified as an inflammation, periapical periodontitis or an acute periapical abscess with pus formation relieved by drainage through the root canal and bone of the jaw (Hillson 1996, 285). In the recorded sample the lesions were either related to caries or occlusal wear and were an external, internal or maxillary sinus drain.

A third of the females and nearly a quarter of the males suffered with abscesses (Table 20). The molars and premolars were the most affected, all to a similar degree. No abscesses were noted in the front teeth (Table 19). The teeth of the mandible were marginally less affected than those of the upper jaw.

Ante-mortem tooth loss

In males and females 34.4% of the permanent maxillary and mandibular teeth were lost ante-mortem. Not surprisingly, the first molar was the most commonly lost tooth, followed by the second and third molars (Table 19).

Dental health: discussion

Dental health in the children

There were 33 subadults in this assemblage and the dental diseases seen in the deciduous teeth were not unusually severe or marked. The number of tooth positions and teeth available to assess was low and this should be considered when looking at the results. Of the 190 tooth positions available 74 had observable erupted teeth. Of the observable teeth evaluated only three (4.1%) had caries and calculus was found to have affected only 17 (23.0%) teeth. Enamel hypoplastic defects when observed were found predominantly on the lower canines. No abscesses or peridontal disease were found in relation to the dentition of the subadults.

One subadult [230] aged at approximately 11 years had particularly marked linear hypoplastic enamel defects of the permanent teeth in the mixed dentition. This individual also manifested healed rickets of the lower limbs, and it is possible that a metabolic disorder may have contributed to the disruption of crown formation in the teeth.

Overall dental health

The overall dental health of this assemblage appears to have been relatively good and particularly if compared to Cross Bones burial ground, where dental health was poor and tooth decay common (Brickley and Miles 1999, 34–6).

For an indication of oral hygiene the high incidence of calculus is probably the most noticeable in the assemblage. In living populations males have more supragingival calculus than females, and the frequency and extent of deposits increase with age (Beiswanger et al 1989, 55–8). This may account for the extent of periodontal disease observed in the assemblage from Chelsea Old Church, which has a relatively high proportion of individuals in the older age categories. The age of some individuals may also have had a bearing on the incidence of caries and tooth loss. Interestingly, in the maxillary teeth the premolars were most affected, whereas in living populations the most frequently affected teeth appear to be the molars (Hillson 1996, 280).

The higher rate of enamel hypoplasia in the males might be due to several factors such as diet, ability to cope with stress and susceptibility to illness (Roberts and Manchester 1995, 58–61).

No dentures or dental interventions or procedures were found in the assemblage. This is interesting as from other broadly contemporaneous cemeteries, such as St Marylebone (Powers 2005, 8) and Christ Church Spitalfields (Whittaker 1993, 53–60), dentures and dental procedures were found with some individuals.

An indication of clay pipe smoking was identified in Richard Butler [198] and one other male [532]. Both manifested the classic rounded pipe facets of the anterior dentition from the habit of holding the pipe between the teeth.

Dental health in the subsample

The make-up of the subsample was such that the two subadults had no teeth present to observe, the 14 males were all over 60 years apart from Richard Butler [198] and the nine females were all over 50 apart from Charity Adams [990].

As the majority were older individuals tooth loss was high

and this may also have affected the incidence of caries recorded. Carious lesions were seen in Thomas Long [654], Catherine Long [722], Edward Rainbows [976], Charity Adams [990] and Collon [1051]. Calculus when present was generally slight apart from Thomas Langfield [147], which was medium to heavy and in the case of Catherine Long [722] who had high ante-mortem tooth loss and heavy calculus. The changes seen particularly on the left side of her mandible might suggest that movement was hindered, which in turn might indicate that she had suffered a stroke.

Nicholas Adams [701], who was 78 years old, had the most severe periodontal disease, and 84-year-old William Wood [681], was completely toothless (Fig 49). Abscesses were found in Milborough Maxwell [792] and Charity Adams [990]. Six individuals had enamel hypoplasia, which was most severe in Richard Butler [198] and Thomas Robson [258], possibly due to serious illness in childhood. Overall the dental health of the named individuals was similar to the remaining individuals in the assemblage.

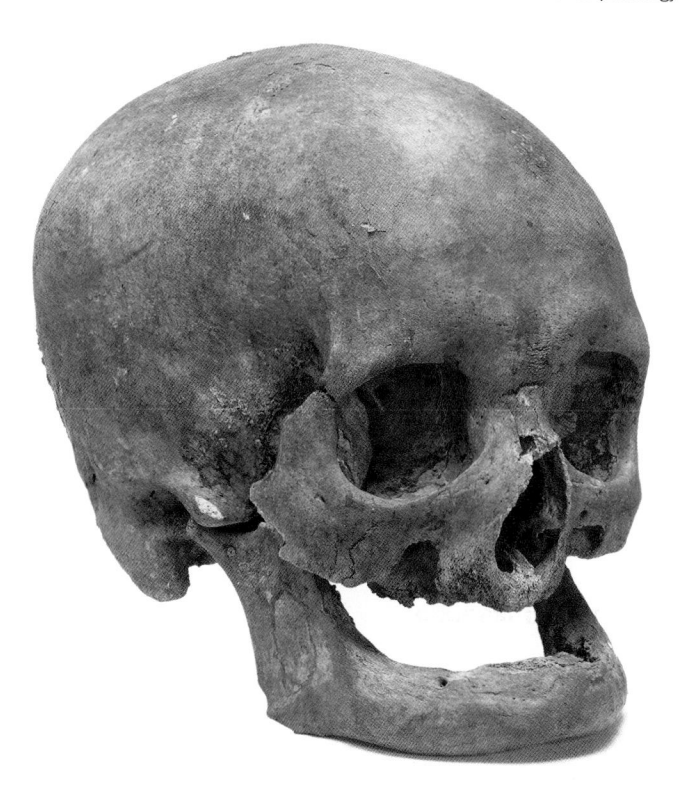

Fig 49 Skull of William Wood [681] (scale c 1:2)

5

Conclusions

The study of Chelsea's past goes back at least 300 years to the time of antiquarians such as John Bowack (1705, 1–13). However, until a few years ago the principal sources of information on this subject were documents, pictures, maps, and standing buildings and monuments. Archaeological evidence was almost non-existent apart from rare chance finds of artefacts, most of which were dredged from the adjacent stretch of the Thames and were without context. Consequently, the periods before c 1500 generally received cursory treatment in local histories.

Since the mid 1990s archaeological excavations in the historic centre of Chelsea and a survey of the adjacent foreshore have begun to address this problem. The archaeological investigation at 2–4 Old Church Street in particular has made a significant contribution to our knowledge of Chelsea before c 1500. It has revealed evidence for prehistoric activity, a Roman rural settlement, and gardens and domestic occupation associated with the medieval manor house. The evidence also sheds new light on the possible origins of the settlement and Old Church Street.

The excavated evidence for post-medieval settlement accords with documentary sources, indicating the presence of great houses in the locality during the 16th and 17th centuries and for domestic occupation of lesser status in the 18th and 19th centuries. Nevertheless, the assemblages of household waste and the evidence recovered from the churchyard have added a new dimension to the study of the social and cultural life of the settlement in the post-medieval period.

The burials from the churchyard are of particular importance because this is the first time that an assemblage from a settlement on the edge of 18th- and 19th-century London has been studied, and the sample has provided valuable demographic data for comparison with groups from London's urban cemeteries. It also provides information about aspects of the health of the community for which there is no documentary evidence. The study of individuals of known age among the burials has confirmed previous studies that indicate that current methods of establishing age-at-death from skeletal remains tend to underestimate the age of older individuals.

Chelsea (and its old parish church) has a distinguished and fascinating history, rivalled by few other former villages in the region, and as yet its store of archaeological data is virtually untapped. However, the little work that has been undertaken so far, especially the excavation at 2–4 Old Church Street, clearly shows the potential for archaeology to investigate aspects of everyday life and death in Chelsea before the modern age.

FRENCH AND GERMAN SUMMARIES

Résumé

Les fouilles archéologiques conduites en 2000 au 2-4 Church Street, à Chelsea, ont révélé des traces d'occupation allant de la Préhistoire jusqu'au XIXe siècle. Le site était localisé immédiatement au nord d'All Saints, la vieille église de Chelsea, qui fut largement reconstruite après le bombardement de 1941.

L'occupation préhistorique était marquée par au moins neuf pièces en silex dont des éclats et un burin mésolithique. Plusieurs structures romaines, incluant des fosses, des fossés et les vestiges éventuels d'un bâtiment en bois rectangulaire constituaient de rares témoins de l'exploitation des environs immédiats de *Londinium*. La principale phase d'occupation semble correspondre au IIIe siècle apr. J.-C.

Aux périodes mediévale et Tudor, le site fut incorporé à l'emprise d'une résidence aristocratique qui était probablement située au nord-est de l'église médiévale. L'activité médiévale la plus ancienne est représentée par une petite quantité de céramique résiduelle datée entre environ 1050 et 1150. Un fossé et plus d'une douzaine de fosses, dont une série sans doute utilisée pour des plantations, étaient probablement associés aux jardins seigneuriaux, principalement aux XIIIe et XIVe siècles. Les matériaux de construction récupérés sur le site étaient fragmentaires et pouvaient, pour partie, provenir de l'église médiévale.

A partir de la fin du XVIIe siècle, la moitié nord du site fut occupée par des maisons ouvrant sur Church Lane (maintenant Old Church Street), avec des jardins et des cours à l'arrière. Les vestiges de deux maisons furent mis au jour. Fosses, puits et latrines parementées en briques, compris entre le XVIe et le XIXe siècle, étaient dispersés dans les anciens jardins. Ces structures ont principalement livré des déchets domestiques et des résidus du jardin, bien que la présence de creusets dans l'une des fosses suggère une petite activité métallurgique à proximité. La découverte de briques rouges « Tudor » et d'éléments luxueux tels que des carreaux de pavement polychromes glaçurés et un carreau de fourneau en faïence s'accordent avec les sources écrites pour établir le statut élevée de cette résidence aux périodes Tudor et Stuart.

A la fin du XVIIe siècle, la partie sud-est du site fut fortement exploitée comme carrière de sable puis un mur en briques fut construit au milieu du terrain pour marquer la limite nord du cimetière. De la terre fut déversée dans le cimetière du côté sud du mur (y compris dans l'ancienne zone d'extraction) pour élever le niveau du sol. L'usage funéraire commença dans la partie septentrionale du cimetière autour de 1700. Elle reçut au moins dix rangées de tombes appartenant au XVIIIe siècle et à la première moitié du XIXe siècle. Deux rangées étaient incorporées à Petyt House, une école et une salle paroissiale qui furent construites dans l'angle nord-ouest du cimetière en 1707. De nombreuses tombes contenaient plusieurs sépultures superposées. On a également observé deux caveaux et deux tombes construites en briques. La plupart des cercueils étaient en bois, à l'exception de neuf cercueils en plomb. Dix-neuf sépultures furent identifiées à partir des plaques apposées sur les cercueils, dont deux membres de la famille Hand qui détenait la célèbre « Chelsea Bun House » (une pâtisserie-confiserie). Deux autres individus ont été identifiés de manière provisoire.

Les squelettes de 290 personnes, dont deux fœtus, furent mis au jour, dont 198 furent sélectionnés pour un enregistrement détaillé dans une base de données ostéologique (Wellcome Osteological Research Database). La plupart furent choisis en raison de leur bonne conservation mais 25 individus le furent aussi, malgré leur état incomplet, en raison des informations biographiques obtenues à partir des plaques de cercueil ou d'autres sources.

Hommes et femmes étaient représentés presque à part égale dans l'échantillon enregistré, qui ne comportait que 16,7 % de subadultes. Beaucoup d'adultes avaient connu une grande longévité. Parmi les 22 personnes dont l'âge fut établi à partir des plaques de cercueils, 13 avaient 60 ans ou plus et 4 étaient octogénaires. L'étude de ce petit ensemble a éclairé les problèmes méthodologiques que pose la détermination de l'âge du décès à partir des restes osseux, qui tendent à sous-estimer l'âge des personnes les plus vieilles.

Des maladies caractéristiques d'un âge avancé, notamment les maladies articulaires, furent identifiées sur un grand nombre d'individus. L'arthrose était également courante, en particulier aux mains et aux poignets, ainsi qu'aux genoux pour les femmes et aux hanches pour les hommes. Dix individus, principalement des femmes, souffraient apparemment d'ostéoporose et un homme de 82 ans était atteint de la maladie osseuse de Paget. La mélorhéostose vertébrale (ou hyperostose vertébrale engainante) (« DISH »), une maladie souvent associée à un régime alimentaire riche et à l'obésité, fut détectée sur neuf hommes et une femme, tous parmi les plus âgés. Soulignons qu'aucune trace de syphilis, supposée répandue à cette période, ne fut détectée.

Les malformations congénitales étaient rares. Des indicateurs possibles de maladies infantiles ou de la sous-alimentation furent trouvés sur plusieurs subadultes et adultes. Des traces de rachitisme actif furent mises en évidence sur un enfant âgé de 1 à 5 ans et de rachitisme guéri sur un enfant de 11 ans. Les fractures touchaient principalement les hommes et la partie supérieure du corps ; elles étaient en général bien réduites et bien cicatrisées et sans trace d'infection secondaire. Des autopsies avaient été pratiquées sur deux hommes.

Zusammenfassung

Archäologische Ausgrabungen im Jahre 2000 in Old Church Street Nr. 2–4 haben Befunde erbracht, die sich von der prähistorischen Zeit bis ins 19. Jh. erstrecken. Der Fundplatz befand sich direkt nördlich von All Saints (Allerheiligen), Chelsea Old Church, die nach der Bombardierung von 1941 weitgehend wieder aufgebaut wurde.

Die prähistorische Nutzung des Platzes wird durch mindestens neun bearbeitete Feuersteine belegt, die Abfälle und ein mesolithischer Stichel beinhalten. Mehrere römische Befunde stellen mit Gruben, Gräben und den möglichen Resten

eines Holzbaus mit rechteckigem Grundriß einen der wenigen Beweise für die ländliche Besiedlung des Hinterlandes von Londinium dar. Die Hauptnutzung scheint im 3. Jh. n.Chr. stattgefunden zu haben.

Im Mittelater und in der Tudor-Zeit lag die Stelle auf dem Gelände eines Herrenhauses, das sich wahrscheinlich nordöstlich der mittelalterlichen Kirche befand. Die früheste Nutzung des Mittelalters tritt uns in einer kleinen Menge von Streukeramik entgegen, die in die Zeit von ca. 1050 bis 1150 datiert. Ein Graben und mehr als zwei Dutzend Gruben hauptsächlich aus dem 13. und 14. Jh. standen wohl im Zusammenhang mit den herrschaftlichen Gärten und umfaßten eine Reihe von möglicherweise Pflanzgruben. Das Baumaterial der Fundstelle war fragmentarisch, einiges stammt vielleicht aus der mittelalterlichen Kirche.

Vom späten 17. Jh. an wurde die nördliche Hälfte der Fundstelle von Häusern, die mit ihrer Fassade zu Church Lane (heute Old Church Street) hin lagen, und ihren Hintergärten und -höfen eingenommen. Reste zweier Gebäude wurden gefunden. Gruben, Brunnen und ziegelverkleidete Senkgruben aus dem 16. bis 19. Jh. waren über die ehemaligen Hintergärten verstreut. Aus diesen Befunden stammen hauptsächlich Haushalts- und Gartenabfälle, obwohl die Tiegel aus einer Grube hier Feinschmiedearbeit in kleinem Stil vermuten lassen. Das Vorkommen von roten „Tudor"-Ziegeln und Luxusartikeln wie polychrome bleiglasierte Bodenfliesen und eine Ofenkachel stimmt überein mit der schriftlichen Dokumentation von Bauten der Oberschicht während der Tudor- und Stuart-Zeit an dieser Stelle.

Im späten 17. Jh. kam es in der Südostecke der Fundstelle zum intensiven Abbau von Sand. Gegen Ende des Jahrhunderts wurde eine Ziegelmauer mitten durch den Platz gezogen, um die nördliche Begrenzung des Friedhofes zu markieren. Erde wurde auf der Südseite der Friedhofsmauer (inklusive der früheren Sandgrube) abgeladen, um das Bodenniveau anzuheben. Um 1700 begann die Bestattungstätigkeit im nördlichen Teil des Friedhofs. Er enthielt mindestens zehn Reihen von Gräbern aus dem 18. und der ersten Hälfte des 19. Jhs. Zwei Reihen lagen innerhalb von Petyt House (einer Schule und eines Gemeinderatshauses, das in der Nordwestecke des Kirchhofs 1707 errichtet wurde). Viele Gräber bargen aufeinander gestapelte Bestattungen. Es gab außerdem zwei Grüfte und zwei ziegelverkleidete Grabkammern. Mit Ausnahme von neun bleiverkleideten Exemplaren bestanden die meisten Särge aus Holz. Neunzehn Bestattungen konnten mit Hilfe der Sargplaketten identifiziert werden; unter ihnen befanden sich zwei Mitglieder der Familie Hand, die das berühmte „Chelsea Bun House" (eine Konditorei) leitete. Zwei weitere Individuen wurden vorläufig identifiziert.

Insgesamt wurden Skelette von 290 Personen, inklusive zwei Föten, aufgedeckt; 198 davon wurden detailliert bearbeitet und fanden Eingang in die Wellcome Osteological Research Database. Bei den meisten gewählten Skeletten stellte ihre relative Vollständigkeit das Auswahlkriterium dar; 25 Individuen wurden allerdings ungeachtet ihrer Erhaltung wegen der biographischen Informationen von den Sargplaketten und anderen Quellen berücksichtigt.

Bei den untersuchten Skeletten gab es annähernd gleich viele Männer und Frauen; nur 16,7 % der Individuen waren nicht erwachsen. Viele der Erwachsenen hatten ein hohes Lebensalter erreicht. Von 22 Individuen, deren Alter den Sargplaketten entnommen werden konnte, waren 13 mindestens 60 Jahre geworden, vier waren als Achtzigjährige verstorben. Die Untersuchung dieser kleinen Stichprobe macht Probleme der derzeitigen Altersbestimmung an Skelettmaterial deutlich, die dazu tendiert, das Sterbealter älterer Personen zu unterschätzen.

Typische Alterserscheinungen, besonders Gelenkkrankheiten, wurden bei einer beträchtlichen Anzahl von Individuen festgestellt. Arthrose war ebenfalls üblich, besonders in den Händen und Handgelenken, in den Knien bei Frauen und den Hüften bei Männern. Zehn Personen, meist Frauen, litten offensichtlich unter Osteoporose, und ein 82-jähriger Mann an der Paget-Krankheit. Hyperostose ankylosans vertebralis („DISH"), eine Krankheit, die oft mit reichhaltiger Ernährung und Fettleibigkeit verbunden wird, wurde bei neun Männern und Frauen höheren Alters diagnostiziert. Interessanterweise gab es keine Hinweise auf Syphilis, die angeblich während dieser Zeit weit verbreitet war.

Angeborene Mißbildungen traten selten auf. Mögliche Indikatoren für Kinderkrankheiten oder Unterernährung wurden bei einigen nichterwachsenen und erwachsenen Personen angetroffen. Symptome für aktive Rachitis gab es bei einem Kind im Alter von 1 bis 5 Jahren, für ausgeheilte Rachitis bei einem 11-jährigen Individuum. Von Knochenbrüchen waren hauptsächlich Männer betroffen, und zwar im Bereich des Oberkörpers. Die Brüche waren im allgemeinen gut geschient und verheilt und wiesen keine Anzeichen für Sekundärinfektionen auf. An zwei Männern fanden sich Spuren von Obduktionen.

BIBLIOGRAPHY

Manuscript sources

Guildhall Library, City of London

ARCHIVES DEPARTMENT
Sun Fire Office Policy no. 439073, 1781

PRINT AND DRAWINGS DEPARTMENT
Engraving by J Roberts of a picture by Jean Baptiste Claude
    Chatelain, c 1750
View of Cheyne Walk from the churchyard of Chelsea Old
    Church by William Parrott, 1840

Kensington and Chelsea Libraries, Local Archives Department, London

SR60 parish sexton's records, 1760–70

DRAWINGS
Drawing of Petyt House, by W W Burgess, c 1890–1
The trade card of Richard Hand of the Chelsea Bun House, 1718
View of the Chelsea Bun House, c 1839

The National Archives (TNA): Public Record Office (PRO)

LEGACY DUTY
IR 26 death duty registers 1796–1903
IR 27 death duty registers, yearly indexes

PREROGATIVE COURT OF CANTERBURY WILLS
PROB 11/530 will of Robert Butler, 1712
PROB 11/696 will of Martha Butler, 1739
PROB 11/1486 will of Thomas Langfield, 1808
PROB 11/1726 will of Nicholas Adams, 1827
PROB 11/1958 will of William Wood, 1835

Victoria and Albert Museum, Department of Design, Prints and Drawings, London (V&A)

E997 to E1011–1903 (M 63e) 1783 trade catalogue of 'J.B.'
E994 to E1021–1978 c 1821–4 trade catalogue by 'A.T.'
E3096 to E3132–1910 1826 trade catalogue by 'E.L.'

Printed and other secondary works

Atkinson, D R, and Oswald, A, 1969 London clay tobacco pipes,
    J Brit Archaeol Ass 32, 171–227
Aufderheide, A C, and Rodriguez-Martin, C, 1998 The Cambridge
    encyclopaedia of human palaeopathology, Cambridge
Barnes, E, 1994 Developmental defects of the axial skeleton in
    palaeopathology, Colorado
Barton, N, 1992 The lost rivers of London, London
Beaver, A, 1892 Memorials of old Chelsea: a new history of the village of
    palaces, London
Beiswanger, B B, Segreto, V A, Mallatt, M E, and Pfeiffer, H J,
    1989 The prevalence and incidence of dental calculus in

adults, J Clin Dentistry 1, 55–8

Bekvalac, J, and Kausmally, T, in prep Chelsea Old Church: an assessment of the accuracy of osteological age and sex determination methods tested on a sample of named individuals from Chelsea Old Church with biographical data, in Proceedings of the seventh annual British association for biological anthropology and osteoarchaeology conference (eds M Clegg, W White and S Zakrzewski), BAR Int Ser, York

Blackmore, L, 1997 The pottery from Cheyne Hospital, Cheyne Walk, Chelsea (CHY96), in Partridge 1997, 27–35

Blackmore, L, 2004 Birdpots, in Sloane, B, and Malcolm, G, Excavations at the Order of the Hospital of St John of Jerusalem, Clerkenwell, London, MoLAS Monogr Ser 20, 276–7, London

Blackmore, L, in prep The pottery, in Pitt, K, with Taylor, J, in prep Finsbury's moated manor house, medieval land use and later development in the Moorfields area, Islington, MoLAS Archaeol Stud Ser

Blunt, R, 1921 By Chelsea Reach: some riverside records, London

Bowack, J, 1705 Antiquities of Middlesex: being a collection of several church monuments in that county, London

Bradley, R, 1990 The passage of arms: an archaeological analysis of prehistoric hoards and votive deposits, Cambridge

Bradley, R, and Gordon, K, 1988 Human skulls from the River Thames, their dating and significance, Antiquity 62, 501–9

Brickley, M, and Agarwal, S, 2003 Techniques for the investigation of age-related bone loss and osteoporosis in archaeological bone, in Bone loss and osteoporosis: an anthropological perspective (eds S Agarwal and S Stout), 157–68, New York

Brickley, M, and Ives, R, 2006 Skeletal manifestations of infantile scurvy, American J Phys Anthropol 129, 163–72

Brickley, M, and Miles, A, with Stainer, H, 1999 The Cross Bones burial ground, Redcross Way, Southwark London, MoLAS Monogr Ser 3, London

British Geological Survey, 1981 England and Wales: south London, 1:50,000, sheet 270, solid and drift geology, Nottingham

Brooks, S T, and Suchey, J M, 1990 Skeletal age determination based on the os pubis: a comparison of the Ascadi-Nemeskeri and Suchey-Brooks methods, J Hum Evol 5, 227–38

Brothwell, D R, 1981 (1963) Digging up bones: the excavation, treatment and study of human skeletal remains, 3 edn, London

Buikstra, J E, and Ubelaker, D H (eds), 1994 Standards for data collection from human skeletal remains: proceedings of a seminar at the Field Museum of Natural History, Arkansas Archaeol Survey Res Ser 44, Indianapolis

The Builder, 1901 2 November, no page numbers

Chambers, R, 1816 Antiquities in the church of Saint Luke, Chelsea, in the county of Middlesex, unpub MS report (copy available in Kensington Public Library)

Cohen, N, in prep Middle Saxon fishtraps: Chelsea, in Cowie, R, and Blackmore, L, Early and Middle Saxon rural settlement in the London region, MoLAS Monogr Ser

Conheeney, J, and Waldron, T, 2002 The human bone from St Bride's lower churchyard, Farringdon Street, London (FAO90), unpub MoL rep

Connell, B, and Rauxloh, P, 2003 A rapid method for recording human skeletal data, unpub MoL rep

Cotter, J P, 1992 The mystery of the Hessian wares: post-medieval triangular crucibles, in Everyday and exotic pottery from Europe c 650–1900 (eds D Gaimster and M Redknap), 256–72, Oxford

Cotton, J F, 1999 Ballast-heavers and battle-axes: the 'Golden Age' of Thames finds, in Mark Dion: archaeology (eds A Coles and M Dion), 58–71, London

Cowie, R, 2002 2–4 Old Church Street, Chelsea, SW3, Royal Borough of Kensington and Chelsea: a post-excavation assessment and updated project design (OCU00), unpub MoL rep

Cox, M, 1996 Life and death in Spitalfields 1700–1850, York

Cox, M (ed), 1998 Grave concerns: death and burial in England 1700–1850, CBA Res Rep 113, York

Crawford Adams, J, 2001 Outline of orthopaedics, Edinburgh

Cubitt, C, 1995 Anglo-Saxon church councils c 650–c 850, London

Cuming, H S, 1857 On the discovery of Celtic crania in the vicinity of London, J Brit Archaeol Ass 13, 237–41

Davies, R, 1904 Chelsea Old Church, London

Denny, B, 1996 Chelsea past, London

Dyer, C, 1994 Everyday life in medieval England, London

Eames, E S, 1980 Catalogue of medieval lead-glazed tiles in the Department of Medieval and Later Antiquities, British Museum (2 vols), London

Egan, G, 2005 Material culture in London in the age of transition, Tudor and Stuart period finds c 1450–c 1700 from excavations at riverside sites in Southwark, MoLAS Monogr Ser 19, London

Egan, G, and Pritchard, F, 1991 Dress accessories c 1150–c 1450, HMSO Medieval Finds Excav London 3, London

Ellis, B M A, 1995 Spurs and spur fittings, in The medieval horse and its equipment, c 1150–c 1450 (ed J Clark), HMSO Medieval Finds Excav London 5, 125–56, London

Farid, S, 2000 An excavation at 6–16 Old Church Street, Royal Borough of Kensington and Chelsea, Trans London Middlesex Archaeol Soc 51, 115–41

Faulkner, T, 1829 Chelsea and its environs: interspersed with biographical anecdotes of illustrious and eminent persons who have resided in Chelsea during the three preceding centuries (2 vols), London

Ferembach, D, Schwidetzky, I, and Stoukal, M, 1980 Recommendations for age and sex diagnoses of skeletons, J Hum Evol 9, 517–49

Gover, J E B, Mawer, A, and Stenton, F M, 1942 The place-names of Middlesex, apart from the City of London, Engl Place-Name Soc 18, Cambridge

Gustafson, G, and Koch, G, 1974 Age estimation up to 16 years of age based on dental development, Odontologisk Revy 25, 297–306

Herring, D A, Saunders, S R, and Katzenberg, M A, 1998 Investigating the weaning process in past populations, American J Phys Anthropol 105, 425–39

Hillson, S, 1996 Dental anthropology, Cambridge

Hohler, C, 1942 Medieval paving tiles in Buckinghamshire, Rec Buckinghamshire 14, 99–132, 149

Holmes, I, 1896 The London burial grounds: notes on their history from the earliest times to the present day, London

Howe, E, 1998 A Romano-British farmstead at St Mary Abbots

Hospital, Marloes Road, Kensington, *Trans London Middlesex Archaeol Soc* 49, 15–30

IGI International genealogical index, http://www.FamilySearch.org

Iscan, M Y, Loth, S R, and Wright, R K, 1984 Age estimation from the rib by phase analysis: white males, *J Forensic Sci* 29, 1094–104

Iscan, M Y, Loth, S R, and Wright, R K, 1985 Age estimation from the rib by phase analysis: white females, *J Forensic Sci* 30, 853–63

Ives, R, 2005 Vitamin D deficiency osteomalacia in a historic urban collection: an investigation of age, sex and lifestyle-related variables, *Palaeopathol Newslett* 130, 6–15

Ives, R, in prep Metabolic bone disease and cortical bone dynamics in British post-medieval urban collections, unpub PhD thesis, University of Birmingham

Janaway, R C, 1993 The textiles, in Reeve and Adams 1993, 93–119

Jarrett, C, 2000a The medieval and later pottery, in Farid 2000, 138–9

Jarrett, C, with Moore, P, Riddler, I, and Farid, S, 2000b The post-medieval pit group [97], in Farid 2000, 126–35

Lambert, P M, 2002 Rib lesions in a prehistoric Puebloan sample from Colorado, *American J Phys Anthropol* 117, 281–92

Lawrence, G F, 1929 Antiquities from the middle Thames, *Archaeol J* 86, 69–98

L'Estange, Rev A G, 1880 *Village of palaces: or, chronicles of Chelsea* (2 vols), London

Litten, J, 1991 *The English way of death: the common funeral since 1450*, London

Longford, E, 1980 *Images of Chelsea*, Richmond-Upon-Thames

Lovejoy, C O, Meindl, R S, Pryzbeck, T R, and Mensforth, R P, 1985 Chronological metamorphosis of the auricular surface of the ilium: a new method for the determination of adult skeletal age at death, *American J Phys Anthropol* 68, 15–28

Lysons, D, 1795 *Environs of London: Vol 2, County of Middlesex*, London

Maresh, M M, 1970 Measurements from roentgenograms, in *Human growth and development* (ed R W McCammon), 157–200, Illinois

Margeson, S, 1993 *Norwich households: the medieval and post-medieval finds from Norwich survey excavations 1971–8*, E Anglian Archaeol Rep 58, Norwich

Matthews, L, and Bell, M, 1957 *Chelsea Old Church*, Fulham

Mays, S, 1998 *The archaeology of human bones*, London

Mays, S, Fysh, E, and Taylor, G M, 2002 Investigation of the link between visceral surface rib lesions and tuberculosis in a medieval skeletal series from England using ancient DNA, *American J Phys Anthropol* 119, 27–86

Mays, S, and Sidell, J, 2003 Head of the river, *Archaeol Matters*, Museum London Newslett (Winter), no page numbers

Mellor, H, 1985 *London cemeteries: an illustrated guide and gazetteer*, Amersham

Merrifield, R, 1983 *London, city of the Romans*, London

Miles, A, 1994 St Bride's lower churchyard: an archaeological excavation (FAO90), unpub MoL rep

Miles, A, 1997 New Bunhill Fields burial ground, Islington

Green: an archaeological watching brief (IGN96), unpub MoL rep

Miles, A, 2002 St Andrew Holborn crypt: an archaeological watching brief (HUD01), unpub MoL rep

Miles, A, and White, W, with Tankard, D, in prep *Burial at the site of the parish church of St Benet Sherehog before and after the Great Fire: excavations at 1 Poultry, City of London*, MoLAS Monogr Ser

Molleson, T, and Cox, M, 1993 *The Spitalfields project: Vol 2, The anthropology: the middling sort*, CBA Res Rep 86, York

Moore, P, 2000 The small finds, in Farid 2000, 134–5

Morris, J (ed), 1975 *Domesday book: 11, Middlesex*, Chichester

Nakhoda, K, and Greene, G, 2005 Diffuse idiopathic skeletal hyperostosis, htpp://www.emedicine.com/radio/topic218.htm

Ortner, D J, 2003 (1981) *Identification of pathological conditions in human skeletal remains*, 2 edn, London

Oswald, A, 1975 *Clay pipes for the archaeologist*, BAR Brit Ser 14, Oxford

Partridge, J, 1997 The Cheyne Hospital, 61–62 Cheyne Walk, SW3, Royal Borough of Kensington and Chelsea: an archaeological post-excavation assessment (CHY96), unpub MoL rep

Perring, D, and Brigham, T, 2000 Londinium and its hinterland: the Roman period, in *The archaeology of Greater London: an assessment of archaeological evidence for human presence in the area now covered by Greater London* (ed MoLAS), MoLAS Monogr, 119–70, London

Phenice, T W, 1969 A newly developed visual method of sexing the os pubis, *American J Phys Anthropol* 30, 297–302

Plume, S, c 1910 *Coffins and coffin making*, London

Powers, N, 2005 Assessment of human remains excavated from St Marylebone School (MAL92 and MBH04), unpub MoL rep

Powers, N, and White, W, in prep Burial and population, in Emery, P, and Wooldridge, K, *Channel Tunnel rail link, new London terminus: archaeological investigation at St Pancras burial ground, 2002–3*, PCA Monogr Ser

RCHM(E), 1925 Roy Comm Hist Monuments (Engl), *An inventory of the historical monuments in London: Vol 2, West London*, London

RCHM(E), 1928 Roy Comm Hist Monuments (Engl), *An inventory of the historical monuments in London: Vol 3, Roman London*, London

Reeve, J, 1998 A view from the metropolis: post-medieval burials in London, in Cox 1998, 213–37

Reeve, J, and Adams, M, 1993 *The Spitalfields project: Vol 1, The archaeology: across the Styx*, CBA Res Rep 85, York

Roberts, C A, and Connell, B, 2004 Palaeopathology, in *Guidelines to the standards for recording human remains* (ed M Brickley), Inst Fld Archaeol Pap 7, 34–9, Reading

Roberts, C A, and Cox, M, 2003 *Health and disease in Britain from prehistory to the present day*, Stroud

Roberts, C A, and Manchester, K, 1995 (1983) *The archaeology of disease*, 2 edn, Stroud

Rogers, J, and Waldron, T, 1995 *A field guide to joint disease in archaeology*, Chichester

Russett, A, and Pocock, T, 2004 *A history of Chelsea Old Church: the church that refused to die*, London

Salter, R B, 1999 (1983) *Textbook of disorders and injuries of the musculoskeletal system*, 3 edn, London

Santos, A L, and Roberts, C A, 2001 A picture of tuberculosis in young Portuguese people in the early 20th century, a multidisciplinary study, *American J Phys Anthropol* 115, 38–49

Scheuer, L, 1998 Age at death and cause of death of the people buried in St Bride's church, Fleet Street, London, in Cox 1998, 100–9

Scheuer, L, and Black, S, 2000 *Developmental juvenile osteology*, London

Scheuer, L, Musgrave, J H, and Evans, S P, 1980 The estimation of late foetal and perinatal age from limb bone length by linear and logarithmic regression, *Annals Hum Biol* 7(3), 257–65

Sisman, A J, and Hochberg, M C, 1993 *Epidemiology of rheumatic diseases*, London

Smith, B H, 1991 Standards of human tooth formation and dental age assessment, in *Advances in dental anthropology* (eds M A Kelly and C S Larsen), 143–68, New York

Stephen, L, and Lee, S (eds), 1908 *Dictionary of national biography*, London

Stewart, Rev W H, 1932 *Chelsea Old Church: an illustrated guide to the parish chapel*, Oxford

Stuart-Macadam, P L, 1989 Nutritional deficiency diseases: a survey of scurvy, rickets and iron deficiency anaemia, in *Reconstruction of life from the skeleton* (eds M Y Iscan and K A R Kennedy), 201–22, New York

Survey of London, 1909 *The parish of Chelsea: Vol 2, Part 1*, London

Survey of London, 1913 *The parish of Chelsea: Vol 4, Part 2*, London

Survey of London, 1921 *The parish of Chelsea: Vol 7, Part 3*, London

Thomas, C, 2003 *London's archaeological secrets: a world city revealed*, London

Trotter, M, 1970 Estimation of stature from intact limb bones, in *Personal identification in mass disasters* (ed T D Stewart), 71–83, Washington

Trotter, M, and Gleser, C G, 1958 A re-evaluation of estimation based on measurements of stature taken during life and of long bones after death, *American J Phys Anthropol* 16, 79–123

Turner, J, 1838 *Burial fees of the principal churches, chapels and new burial grounds in London and its environs ... and all ... information for undertakers*, London

VCH, 2004 *The Victoria history of the county of Middlesex: Vol 12, Chelsea* (ed P E C Croot), Woodbridge

Waldron, T, 1993 The health of the adults, in Molleson and Cox 1993, 67–89

Webber, M, with Ganiaris, H, 2004 The Chelsea club: a Neolithic wooden artefact from the River Thames in London, in *Towards a new Stone Age: aspects of the Neolithic in south-east England* (eds J Cotton and D Field), CBA Res Rep 137, 124–7, York

White, W, 1987 The human skeletal remains from the Broadgate site (WFG62), unpub MoL rep

White, W, 2000 The human remains from the burial ground of St Benet Sherehog (ONE94), unpub MoL rep

Whittaker, D K, 1993 Oral health, in Molleson and Cox 1993, 49–63

Wohl, A S, 1983 *Endangered lives: public health in Victorian Britain*, London

Wroe-Brown, R, 2001 St Paul's Cathedral choir practice facilities: a post-excavation assessment and updated project design, unpub MoL rep

INDEX

Compiled by Susanne Atkin

Page numbers in **bold** indicate illustrations
All street names and locations are in London unless specified otherwise
County names within parentheses refer to historic counties